The Cosmic Constants

A Theory of the Universe

The Cosmic Constants

A Theory of the Universe

Mathew David Hendricks

authorHOUSE®

AuthorHouse™
1663 Liberty Drive
Bloomington, IN 47403
www.authorhouse.com
Phone: 1-800-839-8640

First published by AuthorHouse   09/13/2011

ISBN: 978-1-4567-6132-5 (sc)
ISBN: 978-1-4567-6134-9 (hc)
ISBN: 978-1-4567-6133-2 (ebk)

Library of Congress Control Number: 2011907996

Printed in the United States of America

Table of Contents

Preface

Our universe is a slow motion matter "**Factory**". The matter that our universe continuously produces, is made from the matter that is continuously "emitted" into our universe, at the very center of our universe. This matter contains all of the internal information necessary to produce a larger version of it's smaller self. As this matter travels from the center of our universe "outward", it goes through a specific, continuously balancing amount of accumulation transformations. All of these accumulation transformations happen in a very specific, continuously balancing sequence. Our universe "is" the continuously balancing activity of this entire specific, continuously balancing sequence. All of the continuously balancing behaviors of our universe are "**Constant**". All of the continuously balancing behaviors of our universe are happening "**Now**". The matter that is passing through our universe is on a journey. All matter is on a journey. "One" specific journey. All of matter's behaviors are simply transformations of this "one" specific journey.

This is not a book of answers. This is a book of observations, based solely on the process of elimination.

The Cosmic Constants

"**Everything**" in our universe is produced by different accumulations of atomic matter. Atomic matter consists of very small particles of subatomic matter. These particles, in turn, are made up of even smaller particles of super subatomic matter. These particles, in turn, are made up of even smaller particles of super super subatomic matter, and so on, and so on, and so on....................

Some of the smallest subatomic particles that science has measured are Fermions, Gague Bosons, and Leptons. Fermions are the "moving parts" of every proton, and every neutron. Fermions are held together by Gluons, a form of Gague Boson. Electrons are part of the Lepton group of elementary particles.

One of the smallest subatomic particles that science has measured is a "Quark". A quark is a fermion, and is a component in every proton, and every neutron. There are six types of quarks. Each proton has three quarks. One "down" quark, and two "up" quarks. Each neutron has three quarks, one up quark, and two down quarks. These, and many others, are the continuously moving parts that make up every proton, electron, and neutron. These subatomic particles by themselves are not stable in our universe. They are only stable as a part of an atomic particle. This book recognizes every photon, and electron, and proton, and neutron as "atomic" particles, because that's what atomic matter is made of. Once an atomic particle is separated, the subatomic particles begin their journey back to the center of our universe. If these subatomic particles cannot be employed in our universe, they simply return back to the subatomic universe to be reprocessed back into the first matter of our universe. The first matter, or first particle of our universe is the atomic "neutron". When an atomic neutron exhibits beta decay, it becomes a proton, an electron, and an anti-neutrino. If an anti-neutrino cannot be employed in our universe, it simply returns back to the subatomic universe. This is the "Ucleous". It's the center of our universe.

Everything in our universe is an accumulation of one, or more, of these three atomic particles. The proton, the neutron, and the electron. These "three" base components of atomic matter move very fast, and they are very small. The largest is the neutron.

A "neutron" has no electrical charge, and has a mass of $1,6929$ x 10^{-27} kg.

A "proton" has a positive charge, and has a mass of $1,6726$ x 10^{-27} kg.

An "electron" has a negative charge, and is the smallest of the three, with a mass of only 9.11 x 10^{-31} kg.

Every proton, neutron, and electron possesses an intrinsic mechanical property known as "**Spin**". It's like an object spinning around it's center of mass. "**Spin**" is measured in units of the "Reduced Planck Constant".

Every proton, neutron, and electron all have a "**Spin**" ½. When an electron is in motion around the nucleus of an atom, it also possesses orbital momentum in addition to it's "spin". When a proton, and a neutron form an atom nucleus, this nucleus also possesses angular momentum, due to it's nuclear "spin". The magnetic field of an atom is determined by these forms of angular momentum.

Each of these three atomic particles are "continuously" in motion. All of these continuous motions produce the electric, and magnetism of atomic matter. Every proton, neutron, and electron is a "**Replicant**".

Everything in our universe is an accumulation result of one, or more, of three base atomic particles. Every atom, every molecule, every cell, every rock, every ant, every dog, every human, every tree, every planet, every star, and every galaxy. These are also "**Replicants**". Each replicant is produced in a specific, continuously balancing amount of sequentially compelled environments. Each specific, continuously balancing environment includes a specific, continuously balancing amount of heat, and a specific, continuously balancing amount of pressure.

Our universe is not an electron, or a proton, or a neutron of a larger universe. Our universe is a factory that processes the components necessary to continuously, indefinitely produce the neutron for the next advancing larger universe. This dynamic is identical for every universe. In every universe, smaller and faster matter accumulates in a specific, continuously balancing amount, and a specific, continuously balancing sequence of specific, continuously balancing environments of encouragement, produced by accumulations of larger and slower matter. Our universe is a sphere. Every universe is a sphere. Our universe is encapsulated, and encouraged by a never ending sequence of larger spherical universes. Our universe also encapsulates, and encourages a never ending sequence of smaller spherical universes. Our universe is "identical" to the subatomic universe that continuously produces the neutron for our universe. Our universe is a specific times bigger, and a specific times slower than it's subatomic equivalent, and a specific times smaller, and a specific times faster than it's post atomic equivalent.

Every universe behaves the same. Each one continuously produces neutrons for a larger universe by continuously processing the smaller neutrons from a smaller universe. Our universe is the "nucleus" of an even larger and slower universe.

Every atomic proton, neutron, and electron is a specific, continuously balancing amount, and a specific, continuously balancing sequence of specific, continuously balancing smaller subatomic particle motions, and even smaller, specific, continuously balancing super subatomic particle motions, and so on, and so on, and so on.................... These specific, continuously balancing smaller subatomic particle motions, are exactly the same as the specific, continuously balancing larger atomic particle motions of every proton, every neutron, and every electron. As these three atomic particles accumulate into the "most" balanced larger atomic accumulations, these specific, continuously balancing motions "slow down" to become slow motion, or suspended animation graphs of their smaller and faster particle motions. The "five" specific, continuously balancing motions of every atomic particle, are exactly the same as the "five" specific, continuously balancing celestial motions of every moon, and every planet, and every solar system, and every galaxy. These "five" specific, continuously balancing motions are: **Rotating**, **Wobbling**, **Flipping**, **Quaking**, and **Orbiting**.

At the very outskirts of every universe, the production of the neutron

"**Never**" stops. When the atomic neutron is separated into unified subatomic particle cluster layers, it produces "all" of our universe's protons, electrons, photons, and every other interactive stable atomic particle.

The complexity of the atomic neutron is very basic. It "**knows**" everything there is to "know", and it's components can last "**indefinitely**". When atomic neutrons separate into smaller atomic protons and electron particles, these particles not only can withstand the tremendous pressure, and temperature of a star, they also "**know**" how to produce a star. These atomic particles also "**know**" how to produce "**you**", because different accumulations of protons, neutrons, and electrons is exactly what you are made of. There is "**NO**" magic atomic particle, or atom, or molecule of matter.

You are simply the activation of the specific activity of a specific, continuously balancing amount, and a specific, continuously balancing sequence of smaller activators. This pattern of matter is constant. Every electron is the activation of the specific activity of a specific, continuously balancing amount, and a specific, continuously balancing sequence of smaller activators, each one of these smaller activators are the activation of the specific activity of a specific, continuously balancing amount, and a specific, continuously balancing sequence of even smaller activators, and so on, and so on, and so on.....................

The battery of very fast, and very small matter is "**Always**" even smaller, and even faster matter. "**Always**". This pattern of matter is constant.

All matter, in every smaller or larger universe, is the same "**age**". **Timeless**.

In the cosmos, there is "no" scale of time, and there is "no" scale of size. When science looks deeper into space, they will "always" find the same thing, "more". When science looks deeper into the subatomic particle, they will "always" find the same thing, "more".

All matter is "three" dimensional. There is nothing that is two dimensional, or four dimensional, or any other dimensional. All atomic matter is a direct energy output percentage, to indirect energy output percentage, in relation to overall energy input, continuously balancing equation.

The atomic neutron is the "product" of the subatomic universe. It is an "organized" compilation of subatomic particle cluster layers. There are "NO" organizers inside of the neutron, the proton, the electron, or the photon. There is only a Deep Torsion Constant of never ending, overlapping, smaller and smaller identical layers of centrifugal, and centripetal, rotating, wobbling, flipping, quaking, and orbiting particle formations. These particle formations are produced and organized in the subatomic universe by advanced subatomic human-like creatures.

When we look out into space, we see galaxys. We are also looking at one of the "exact" same dynamics used to produce the atomic neutron, only a specific amount bigger, and a specific amount slower. Our universe is producing the "identical" product of the subatomic universe by using the "identical" accumulating dynamics. Our universe's neutron is a specific times larger, and a specific times slower than it's atomic equivalent.

Atomic matter became "something" that entered our universe, and it is becoming something that will leave our universe. The only "clues" there are to what this matter was, or what this matter will be, is what this matter is now, and the behaviors that it exhibits. Larger atomic matter "always" does what smaller atomic matter does, only much slower, and in a transformed way. Matter only does one thing. Only one. Matter continuously reproduces itself by using a specific, continuously balancing amount, and a specific, continuously balancing sequence of accumulation transformations of it's smaller self. This specific, continuously balancing amount, and specific, continuously balancing sequence of accumulation transformations is our universe.

Matter exhibits a "maximum complexity retention" dynamic. "Every" universe continuously produces complex components with a long cosmic lifespan, to continuously produce "more" complex components with an even "longer" cosmic lifespan, to continuously produce the "most" complex component with the "longest" cosmic lifespan, which is the cosmic neutron.

In a "specific", continuously balancing environment including a "specific", continuously balancing amount of heat, and a "specific", continuously balancing amount of pressure, matter can turn two protons into an atomic nucleus of one proton, and one neutron. That's because "space" itself is "continuously" interactive with atomic matter. "Space" is comprised of "every" smaller and faster universe's "neutrons" returning back to each of their individual original universes to be reprocessed. A specific amount of "Space's" neutrons are subatomic neutrons returning back to the subatomic universe. These returning subatomic neutrons "continuously" interact with our universe's atomic matter. Atomic neutrons are "continuously" returning back to our universe through our universe's five "**Uniospheres**", from the post atomic universe.

As matter "continuously" accumulates, it "continuously" travels outward from the center of it's original universe. As matter "continuously" disintegrates, it "continuously" travels inward back to it's original universe. All matter is continuously flowing in the cosmic continuum. Larger and slower accumulations of matter continuously flow one way, and smaller and faster disintegrations of matter continuously flow in the other direction. The only clues there are to matter's smaller and faster accumulations, is the continuous behavior of matter's larger and slower accumulations. Larger matter "always" does what smaller matter does. Only much slower, and in a transformed way.

As we follow accumulating atomic matter's journey outward from the ucleous to it's present point of accumulation, matter "itself" reveals "all" of it's intentions by continuously exhibiting continuous, repetitive patterns of behavior. When atomic matter accumulates, it's behavior not only slows down, but it also transforms. "All" of the functional dynamics of atomic matter are "always" present in every atomic cosmic "whole" accumulation. They may not be visible, but they are "always" present. The continuous, repetitive patterns of matter are simply specific transformations of the continuous journey of matter. This is a compilation of some of the continuous, repetitive patterns of atomic matter. This is the "journey" of matter.

The Patterns of Matter

The journey of the atomic neutron begins at the subatomic universe. When an atomic neutron is emitted into our universe, it represents the "End" of the subatomic accumulation transformation sequence cycle, as well as the "Beginning" of the atomic accumulation transformation sequence cycle. This is also the beginning of all the repetitive patterns that accumulated atomic matter exhibits.

For the subatomic universe, the atomic neutron is the maximum complexity retentive equation result. For the atomic universe, the atomic neutron is the first component employed to initiate our universe's maximum complexity retention equation result. Everything matter produces is a replicant. There is a specific, continuously balancing amount of them, and they are continuously produced in a specific, continuously balancing sequence cycle of accumulation. This specific, continuously balancing sequence cycle is our universe. Each larger or smaller successive universe functions identically to our universe. Our universe's specific, continuously balancing amount, and sequence cycle of matter's continuous accumulation transformations begins at the center of our universe. The subatomic universe is the "nucleus" of our universe. This is the "Ucleous".

If we could see the "Ucleous", we would see a rotating, wobbling, quaking, and periodically flipping sphere. The "Ucleous" continuously produces replicant atomic neutrons. The specific, continuously balancing pressure of our universe, and the specific, continuously balancing attractiveness of the "Ucleous", encourages the continuous production of atomic neutrons to accumulate into a unified layer, encapsulating the

"Ucleous". This continuously accumulating unified sphere of atomic neutrons is the "Univator". As the amount of unified atomic neutrons continuously increases, so does the size of the "Univator". At a certain point, the specific, continuously balancing pressure of the universe, and the specific, continuously balancing attraction of the ucleous, stops the univators size from increasing, even though the amount of atomic neutrons being produced by the ucleous "never" stops.

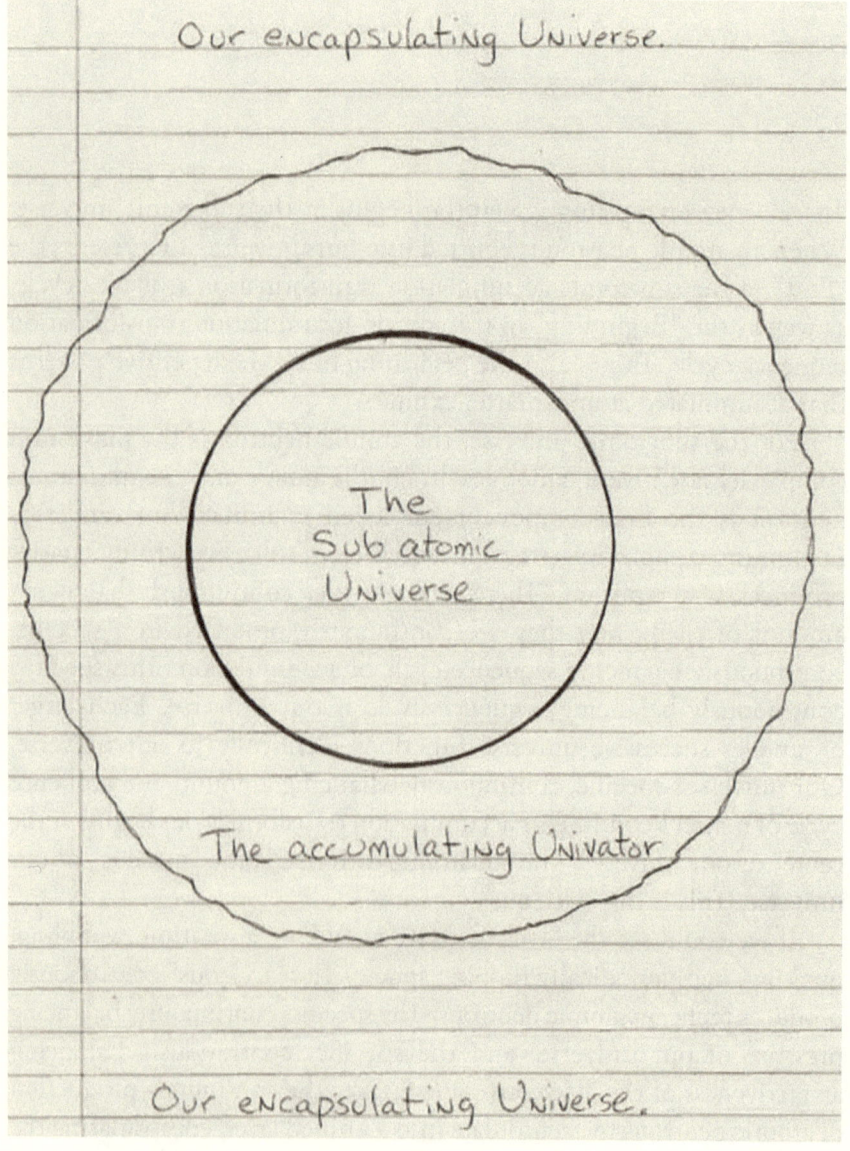

Our encapsulating Universe.

The
Sub atomic
Universe

The accumulating Univator

Our encapsulating Universe.

This is the first of two "rests" in the cosmic lifespan of the univator. After the first "rest", the univator's size will begin to decrease as it collapses around the ucleous. This first "rest" may be one millisecond, or it may be one trillion years. Never the less, after the univator's first rest, it begins to collapse on itself back toward the ucleous. This is the beginning of the "**Univation**", or "The Big Bang". There have been countless big bangs in the past, and they will continue indefinitely. As the univator collapses on itself and the encapsulated ucleous, it's size will decrease until again, it must "rest". This rest also may be one millisecond, or one trillion years. After the second rest is over, there is an explosion of atomic particles. These new specific replicant atomic particles speed away from the ucleous in a spherical "pulse", or "wave". This specific spherical wave of more diverse atomic replicants is "encouraged" to produce the next phase of matter's maximum complexity retention equation.

The next unified layer of unified particle clusters, in the universe's specific, continuously balancing amount, and specific, continuously balancing sequence of unified particle cluster layers, is produced by unified clusters of radiant galaxy activating stars. These super stars are produced by the pulse, or wave of atomic replicants, and a replicant galaxy activator "seed", or "galactivator". Each galactivator is an atomic particle accumulation regulator. Each galactivator is identical. When the radiant state of a galaxy activating super star is over, it will have produced all of the specific atomic replicants that are needed to produce a galaxy, plus an "attractive" rotating, wobbling, flipping, quaking, and orbiting center to "encourage" a galaxy's formation. This more complex environment is where our universe's next unified layer of unified particle clusters begins.

The next unified layer of unified particle clusters is produced by unified semi radiant, or "unified active galaxy clusters". Each galaxy starts off as a replicant. The galaxys we see today are in different phases of accumulation. They "all" start the same, and they "all" transform as they accumulate and interact with each other. Eventually each accumulated galaxy center will be a particle in the next advancing unified particle cluster layer, in the universe's specific, unified particle cluster layer sequence.

Every galaxy is comprised of "many" different replicant stars. Each star produces a more complex environment, as well as producing a more complex specific assortment of atomic particle, and molecular replicants. All of these replicant stars have a very long radiant cosmic life span. "Every" replicant star is used for gyroscopic torsion to "steer", or "guide" the galaxy, but there are only four specific replicant stars, plus the galaxy activating super star, that when fully processed, become the post atomic neutron.

Each one of these "four", post atomic neutron forming stars, are produced by a specific "seed" that regulates the hydrogen, or other atomic particle's accumulation. Each star is encouraged to accumulate while orbiting a fully processed, galaxy activating super star seed. Each star is also encouraged to accumulate by the specific, continuously balancing attraction of the subatomic universe, and the specific continuously balancing repulsion, or pressure of our universe.

The "largest" of these stars is three times the mass of our star, the sun. It's radiant state is between five, and ten million years. When this star's radiant state is over, most of it's mass will collapse on itself producing a very small spinning dark star. It is "very" dense, and "very" attractive. The core of this dark star is a fully processed star seed.

The second "largest" of these stars is two times the mass of the sun. This star's radiant state is also about five, to ten million years. When the radiant state of this star is over, most of it's mass collapses on itself producing a small spinning neutron star. It is very dense, and very attractive. The core of this neutron star is a fully processed star seed.

The third "largest" of these stars is one and a half times the mass of the sun. This star also has a radiant state of about five, to ten million years. When this star's radiant state is over, most of it's mass collapses on itself. At a certain point during this star's collapse, part of this star's mass explodes out into space. The remainder of it's mass continues to collapse on itself producing a small spinning neutron star. When this star explodes, it is a "super nova". This super nova produces a specific, more complex environment, as well as a specific amount, and a specific, more complex assortment of atomic and molecular replicants. This specific, more complex environment, and more complex specific assortment of atomic and molecular replicants is what every replicant "forth star" is produced from. It is also what every replicant "earth" is produced from. The core of this neutron star is a fully processed star seed. It is very dense, and very attractive.

The "smallest" of these four galactic stars is "our" star, the sun. This star's radiant state is ten billion years. When the radiant state of this star is over, most of it's mass will collapse on itself producing a small spinning white dwarf star. It is very dense, and very attractive. The core of this white dwarf star is a fully processed star seed. This star's ten billion years long radiant state is the incubation dynamic for the next sequential higher level of matter's maximum complexity retention. This replicant star provides the specific stabilizing dynamic for "every" replicant earth's accumulation. This replicant star, and it's replicant earth, now produce a specific, even more complex environment, producing a specific amount and sequence of specific, even more complex atomic and molecular replicants. When the environment "changes", so do the replicants. When the environment becomes more and more complex, so do the atomic and molecular replicants.

In our galaxy alone, there are "billions" of the largest radiant star, and "billions" of it's non radiant dark stars. There are "billions" of the second largest radiant star, and "billions" of it's non radiant neutron stars. There are "billions" of the third largest radiant star, and "billions" of it's non radiant neutron stars. There are "billions" of radiant stars that are identical to our sun, and "billions" of it's non radiant white dwarf stars. In our galaxy "alone", there are "billions" of earth replicants, "BILLIONS".

On "every" earth, the geography is not the same, but the accumulation dynamic of atomic and molecular matter is "identical". Every replicant sun's incubation dynamic of a specific, continuously balancing environment including a specific, continuously balancing amount of heat, and a specific, continuously balancing amount of pressure, plus every replicant earth's own incubation dynamic of a specific, continuously balancing environment including a specific, continuously balancing amount of heat, and a specific, continuously balancing amount of pressure "Together", produce the "two" transformed more complex replicants that will produce the next sequential higher level of our galaxy's maximum complexity retention.

Out in "space" there are "many" specifically different molecules, and organic molecules. "All" of them are produced by specific, continuously balancing environments, and the specific, continuously balancing heat and pressure of billions of radiant stars, and billions of super novas.

Four billion years ago, our young earth was saturated with accumulated clouds of gaseous molecular activity. Three of these gasses were hydrogen rich methane, ammonia, and water vapor. When a specific, continuously balancing amount of each of "all" three gasses are "electrified", they produce complex mini proteins, and complex mini nucleic acids that can "reproduce" themselves. This is how "all" matter is transformed. In a specific, continuously balancing amount, and a specific, continuously balancing sequence of specific, continuously balancing environments. Each specific, continuously balancing environment includes a specific, continuously balancing amount of heat, and a specific, continuously balancing amount of pressure.

Every atom, and molecule, and organic molecule, and cell, and plant, and insect, and animal is a sequentially compelled atomic cosmic "whole" component, that is a specific compilation of many smaller cosmic "whole" components existing as "one". Each sequentially compelled atomic cosmic "whole" replicant is produced by a specific, continuously balancing amount, and a specific, continuously balancing sequence of specific, continuously balancing sequentially compelled environments. Each specific, continuously balancing environment includes a specific, continuously balancing amount of heat, and a specific, continuously balancing amount of pressure.

The spacific fifty, or so, organic molecules that produce "every" insect, and "every" plant, and "every" animal on "every" earth are not "magical". They are an equation result of atomic matter's accumulating behavior.

"Every" one of these specific fifty, or so, molecules is an accumulation of twenty-five, or so, specific atoms. Each specific atom is an accumulation of three specific atomic particles. Protons, electrons, and neutrons. There is no "magic" atomic particle of biological matter. There is only specific, continuously balancing amounts of specific, continuously balancing components in specific, continuously balancing sequences.

The evolution of atomic particles becoming biological organisms is the result of a specific, continuously balancing amount, and a specific, continuously balancing sequence of compelled environments. Biological matter is not "magical", it is a transformed overall motion result produced by specific compilation accumulations of three atomic particles.

Biological matter is the result of the "continuous" external encouragement from larger and slower accumulations of atomic matter, **"And"** the "continuous" internal response from smaller and faster accumulations of atomic matter. The evolution of atomic particles becoming biological organisms is "Identical" on "Every" Earth in our milky way galaxy, and "Every" other active galaxy in our universe. Every replicant earth and it's replicant biological organisms may look different, but their overall anatomy and functions are "Identical".

Every Earth in our universe is produced for the exact same reason.

Cosmic Engineering

Every atomic particle is a specific, continuously balancing division of an atomic neutron. Each atomic particle is a specific, continuously balancing accumulation of a specific, continuously balancing amount, and a specific, continuously balancing sequence of specific, continuously balancing components behaving as "one" cosmic whole component. Every neutron is one "Complete", continuously balancing unit of cosmic information. Every proton, electron, and photon is one specific "Incomplete", continuously balancing unit of cosmic information. Each one is a machine with no touching parts.

The atomic "neutron" is the cosmic equation result. One atomic neutron is a machine with no touching parts. One atomic neutron contains the knowledge of the entire subatomic universe. It also contains the knowledge of the entire super subatomic universe, and the super super subatomic universe, and the super super super subatomic universe, and so on, and so on, and so on.....................

One neutron is one "complete", continuously balancing amount, and one "complete", continuously balancing sequence of subatomic information. It is one "complete", continuously balancing "unit" of subatomic information. This "complete", continuously balancing unit of subatomic information encapsulates one "complete", continuously balancing amount, and one "complete", continuously balancing sequence of super subatomic "complete", continuously balancing units, which encapsulates one "complete", continuously balancing amount, and one "complete", continuously balancing sequence of super super subatomic "complete", continuously balancing units, which encapsulates one "complete", continuously balancing amount, and one "complete", continuously balancing sequence of super super super subatomic "complete", continuously balancing units, and so on, and so on, and so on.....................

The atomic "neutron" is continuously "told" what to do by it's specific, continuously balancing accumulation of subatomic components. The specific, continuously balancing accumulation of subatomic components are continuously "told" what to do by their specific, continuously balancing accumulation of super subatomic components, which are continuously "told" what to do by their specific, continuously balancing accumulation of super super subatomic components, which are continuously "told" what to do by their specific, continuously balancing accumulation of super super super subatomic components, and so on, and so on, and so on.....................

The behavior of the atomic neutron is produced by the endless layering of specific, continuously balancing amounts, and specific, continuously balancing sequences of "Identical" smaller and faster information. This "information" is produced by "continuously" moving, interacting, everlasting, centripetal and centrifugal, rotating, wobbling, flipping, quaking, and orbiting particle clusters.

This "Deep Torsion Constant" of information produces the "continuous" request and response behavior of matter. Larger and slower accumulations of matter "always" know what they will request smaller and faster accumulations of matter to do, and smaller and faster accumulations of matter "always" respond accordingly.

Matter "always" knows what it's doing.

Alive

Every proton, electron, and neutron is as "alive" as we are, because that's what we are made of. Your body is made up of approximately one hundred trillion cells. Your cells are "alive". Your cells are alive, because the molecules that produce your cells are alive. Your molecules are alive, because the atoms that produce your molecules are alive. Your atoms are alive, because your protons, electrons, and neutrons are alive. Your protons, electrons, and neutrons are alive, because a continuously balancing accumulation of protons, electrons, and neutrons is "**You**".

There is no "magic" atomic particle of biological matter. There is no "magic" atom, or molecule of biological matter. The specific atoms and molecules that produce biological life, are simply the atoms and molecules that when compiled in specific accumulation compilations, produce specific transformed motion results.

Every proton, electron, and neutron is a smaller and faster transformed motion result of exactly what "**You**" are. Atomic matter only produces what it is. Atomic matter cannot produce what it is "not". All of your functional dynamics are transformed slow motion, and suspended animation "graphs" of "all" of the functional dynamics of smaller and faster atomic matter. "All" of the functional dynamics of biological matter, as well as atomic matter, is a transformation of "**one**" specific system. The system of "continuously balancing information circulation". "All" of matter's "continuously balancing information circulations" are transformations of one specific, continuously balancing motion.

17

Everything that matter produces is a specific, continuously balancing amount, and a specific, continuously balancing sequence of specific transformations of "one" specific, continuously balancing journey of "out half around rest, half around back rest".

As atomic matter accumulates into atomic particles, and atoms, it begins to "slow down". As atomic matter accumulates into molecules and organic molecules, it "slows down" even more. As atomic matter accumulates into the most complex accumulations of atomic and biological matter, like ants, or trees, or humans, or planets, or stars, it slows down enough to reveal "all" of it's smaller and faster functional dynamics. Atomic matter produces "life", because it is "alive". It is simply a different "form" of life in each one of it's specific accumulation transformations. "All" of the continuous behaviors that "you" exhibit, are continuously exhibited, in a transformed way, by "every" atomic cosmic whole component. Every electron "exfoliates". Every atom "eats". Every proton has a "skeletal system". Every star has an "endocrine system". Every cell has a "skin". Every earth has a "respiratory system". Every molecule has a "digestive system". Every photon has an "immune system". Every neutron has a "neural system".

Biological "life" does not produce specifically different organic molecules. Specifically different organic molecules produce biological "life". These specific organic molecules are produced by specific accumulations of specific atoms. These specific atoms are specific accumulations of three atomic particles. Protons, electrons, and neutrons. **There is no** "magic" particle of biological matter. Each specific atomic particle of biological matter is a specific, continuously balancing amount, and a specific, continuously balancing sequence of specific, continuously balancing repetitive subatomic particle "motions". These are produced by a specific, continuously balancing amount, and a specific, continuously balancing sequence of specific, continuously balancing repetitive super subatomic particle "motions", and so on, and so on, and so on.................... These specific, continuously balancing repetititve "motions" produce specific, continuously balancing repetititve "motion results". These specific, continuously balancing repetititve "motion results" are specific, continuously balancing repetititve "**SOUNDS**".

Each specific atom of biological matter is produced in a specific, continuously balancing amount, and a specific, continuously balancing sequence of specific, continuously balancing environments. Each specific, continuously balancing environment includes a specific, continuously balancing amount of heat, and a specific, continuously balancing amount of pressure.

Each specific organic molecule of biological matter is produced in a specific, continuously balancing amount, and a specific, continuously balancing sequence of specific, continuously balancing environments. Each specific, continuously balancing environment includes a specific, continuously balancing amount of heat, and a specific, continuously balancing amount of pressure.

Each specific cell of botanical, or biological matter is produced in a specific, continuously balancing amount, and a specific, continuously balancing sequence of specific, continuously balancing environments. Each specific, continuously balancing environment includes a specific, continuously balancing amount of heat, and a specific, continuously balancing amount of pressure.

"**YOU**" are produced from a specific, continuously balancing amount, and a specific, continuously balancing sequence of specific, continuously balancing environments. Each specific, continuously balancing environment includes a specific, continuously balancing amount of heat, and a specific, continuously balancing amount of pressure.

Biological Molecular Circulation

Every biological organism is produced from "one" specific assortment, and "one" specific amount of base organic molecules. These fifty, or so, specific organic molecules are made up of "one" specific assortment, and "one" specific amount of base atoms. Each specific twenty-five, or so, base atoms are made up of one specific assortment, and one specific amount of three specific atomic particles.

These fifty, or so, spacific organic molecules accumulate into complex compilations in a specific, continuously balancing amount, and a specific, continuously balancing sequence of specific, continuously balancing environments. Each specific, continuously balancing environment includes a specific, continuously balancing amount of heat, and a specific, continuously balancing amount of pressure. Some of these complex compilations of organic molecules are the food we eat, and the air we breathe.

When you eat something, you chew it into smaller pieces and swallow it. These smaller pieces of complex compilations of organic molecules are "**SHAKEN**" apart by your stomach acids. Some of these smaller compilations of specific organic molecules are "escorted" to the area of your body's cell production, and some are continuously escorted to your presently active cells, to continuously circulate in and out of your presently active cells, to keep them continuously, presently active. As new cells accumulate, old cells die. Your body is constantly replacing old cells with new cells at a rate of about five million, to eight million cells per second. Some cells "die" from "wear and tear". Some cells simply reach the end of their cosmic lifespan, and some cells simply self destruct when they are "told" to. The life cycle of "every" cell is carefully monitored and regulated so that you "always" have the right amount of each specific cell. all of your new cells are produced from the food you eat, the liquid you drink, and the air you breathe.

Every time you eat something, you are eating molecules that are "**new**" to your body. Every time you drink something, you are drinking molecules that are "**new**" to your body. Every time you breathe in, you are breathing in molecules that are "**new**" to your body. These "**new**" molecules are what "**new**" cells are made of. "**New**" specific organic molecules are "**continuously**" entering your body.

Your dead cells, or your "**old**" cells, are "**continuously**" exiting your body through exhalation, exfoliation, transpiration and periodic excretion. All of your body's "**old**" cells, which are dead cells, continuously exit your body. These "**old**" cells are made up of millions and millions of "**old**" molecules. "**Old**" specific organic molecules are "continuously" exiting your body.

When your "**old**" cells die, your "**new**" cells replace them. Each "**new**" cell is filled with "**new**" molecules, and each "**old**" cell is filled with "**old**" molecules. The "**new**" specific fifty, or so, organic molecules that continuously enter your body are not "**new**" to our universe, or to our solar system, or to our earth. They are only "**new**" to your present overall cosmic whole molecular compilation. The "**old**" specific organic molecules inside your dead cells that are continuously exiting your body are not dead, and they are not necessarily "**old**" to our universe, or to our solar system, or to our earth. They are only "**old**" to your present overall cosmic whole molecular compilation.

Your body contains about one hundred trillion active cells. Some of your cells live longer than others, but they all exhibit a finite cosmic life span. With a cell death rate of five million cells per second, at the end of one year, your body will have discarded well over one hundred trillion "**old**" cells filled with "**old**" molecules, and replaced them with well over one hundred trillion "**new**" cells filled with "**new**" molecules. Between your continuous cellular circulation, and your continuous molecular circulation, at the end of one year, at the molecular level, you are not the same you as last year. Some of your live teeth cells or live bone cells may still be present, and some of your dead hair cells or dead nail cells may still remain, but molecularly, over ninety persent of the organic molecules that made up your overall cosmic "**whole**" last year, have been replaced with completely "**new**" organic molecules. Over ninty percent of the organic molecules that were part of you last year, are no longer part of you this year. Molecularly, you are not last year's you. Molecularly, you do not have the same heart, or lungs, or liver as last year. Molecularly, you do not have the same blood as last year, or the same skin. Molecularly, you do not have the same brain as last year.

At the organic molecular level, you are "**never**" the same person. Every time you breathe in, your overall organic molecular compilation changes. At the organic molecular level, you are "**never**" the same person twice. If every organic molecule that circulates throughout our biosphere had it's own specifically different visible identification signature, you would be able to "see" that at the organic molecular level, you are "**never**" the same molecular compilation. Your organic molecular "circulation" "**never**" stops. It is continuous. It is constant. "Biologically", you are "always" one cosmic whole component. "Molecularly", you are "**never**" the same compilation. At the organic molecular level, you do not have last year's brain, yet you still can remember last year. That's because your specific organic molecules remember. Your specific organic molecules remember, because your specific atoms remember. Your specific atoms remember, because your specific atomic particles remember, and your specific atomic particles remember, because their specific subatomic particles remember, and your specific subatomic particles remember, because their specific super subatomic particles remember, and so on, and so on, and so on.....................

In the cosmos, there is no such thing as "learning" or "teaching". There is only memory.

Atomic Circulation

At the atomic level, you are "never" the same compilation, or "person". At the atomic level, you are "never" the same compilation, or "person" twice. Your atomic circulation "never" stops. It is continuous. It is constant.

At the atomic level, you have circulated the equivalent of a few truckloads of carbon in and out of your body. You have circulated the equivalent of a small lake in and out of your body. You have circulated enough iron in and out of your body to make a few boxes of nails. The protons, electrons, and neutrons that make up your specific atoms, and your specific organic molecules, have been in our universe at "least" thirteen billion years. They could be older than that. Biologically, you may be twenty, thirty, or forty years old, but at the atomic particle level, you are easily thirteen billion years old, or older. When you touch your forefinger with your thumb, you are "touching" particles that are thirteen billion years old.

"All" of the atomic particles in your body used to be inside of a star. They may have circulated in and out of hundreds, or maybe millions of stars. "All" of the atomic particles in your body used to be inside of a galaxy activating super star. "All" of the atomic particles in your body used to be inside the nucleus of our universe. The "Ucleous". Now, billions and billions of years later, after a specific amount, and a specific sequence of specific accumulation transformations, "all" of your atomic particles are you. There is "**no**" magic atomic particle of biological matter. There is "**no**" magic atom of biological matter. There is "**no**" magic organic molecule of biological matter. There is "**no**" magic cell of biology, or botany. There is only specific atomic accumulations, producing specific transformations. There is "**no**" random behavior in the cosmos, just a "tremendous" amount of expected results produced by the Deep Torsion Constant.

The Brain Of Matter

"Every" atomic particle is a "brain". Every atom is a "brain". Every organic molecule is a brain. Your brain does not produce specific atomic particles, or specific atoms, or specific organic molecules. Specific atomic particle's, and specific atoms, and specific organic molecules produce your brain.

When a specific amount of hydrogen atoms accumulate in space, they become a specific star. Each hydrogen atom is externally encouraged to "accumulate", and each hydrogen atom is internally motivated to respond. Each hydrogen atom responds accordingly. Consistently. Continuously. Constantly. The "Brain" of a star is not it's heat, or size. The brain of a star is it's specific, continuously balancing environment, and it's regulated accumulation of specific, continuously balancing atomic particles. A star accumulates, and "shines" for millions, and sometimes billions of years. This is because it's specific accumulated atoms "**know**" what to do in this specific accumulated state, and they do it.

Your brain is the same as a star's brain. Your brain is your specific, continuously balancing environment, and your regulated accumulation of specific, continuously balancing atomic particles. In between your specific, continuously balancing environment brain, and your specific, continuously balancing atomic particle's brains, is a specific, continuously balancing amount, and a specific, continuously balancing sequence of transformed information translating brain layers, which is you. Your digestive system is a brain. Your skeletal system is a brain.

27

Your immune system is a continuously calculating and responding brain. Every compilation of accumulated matter is a brain.

When you look at your finger, you're looking at a brain. It is a transmitter, a receiver, and a warehouse of information. Your finger brain, and your two pound grey brain, are continuously communicating. It "never" stops. When your grey brain wants to touch something, it employs the finger brain. When the finger brain receives a message of lift, push, pick, slow, fast, strong or delicate, it understands and responds. Your finger brain lets your grey brain know how soft something is, or how hard, or how wet, or how hot, or how cold something is. When your grey brain tells your finger brain to "go here", and "here" is too hot, your finger brain says, "I am here now, but it is too hot to stay". Your grey brain says "move away quickly". Your finger brain receives this message, and it responds. Your grey brain can ignore this message, but your finger brain will only send back the same message. Your finger brain will not say "cold" instead of "hot". Your finger brain knows the difference. To your grey brain, something may "look" cold, but your finger brain cannot be fooled. It "always" knows the difference between hot and cold. Your grey brain's survival depends on it. Your finger brain is itself a specific compilation of brain layers. The "push" and "pull" brains are the muscles, and the "hot" and "cold" brains are the skin. Every inch of your skin is a brain. Every nerve ending on your skin is a brain. Each nerve ending on your skin is a warehouse of information. Each one is continuously sending and receiving complex information to and from your grey brain in a "never ending" circulation of communication. Your whole body is a specific amount, and a specific sequence of transformed information translating brain layers. Each specific brain layer is in constant communication with each and every other specific transformed brain layer. Each specific transformed brain layer sequentially encapsulates the others. Each specific transformed sequential brain layer is the translator between a larger transformed encapsulating brain layer, and a smaller transformed encapsulated brain layer. Just like our Earth's atmosphere encapsulates you. This continuous specific atmospheric information is continuously translated to your encapsulated grey brain, by your encapsulating skin brain. Your grey brain encapsulates a specific amount, and a specific sequence of encapsulated brain layers, which instantaneously translates the grey brains information, and instantaneously tells your grey brain what to tell your skin brain to do. If this atmospheric information is "cold", you instantaneously get goosebumps. If this atmospheric information is "hot", you instantaneously start to sweat. Your skin brain does not communicate directly with your white blood cell brains, or your electron brains, but "all" of them are "always" in constant, continuous communication with each other through a specific amount, and a specific sequence of transformed information translating brain layers. You're encapsulating larger brain layers can only "encourage" your smaller encapsulated brain layers. The smaller encapsulated brain layer is "always" the "brains" behind the larger encapsulating brain layer.

Your skin brain is the "brains" behind your atmospheric brain. Your grey brain is the "brains" behind your skin brain. Your lymbic system is the "brains" behind your grey brain. Your R complex is the "brains" behind your lymbic system. Your brain stem is the "brains" behind you R complex. Your white blood cells are the "brains" behind your brain stem. Your DNA is the "brains" behind your white blood cells. Your fifty, or so, specific organic molecules are the "brains" behind your DNA. Your twenty-five, or so, specific atoms are the "brains" behind your fifty, or so, specific organic molecules. Your three specific atomic particles are the "brains" behind your twenty-five, or so, specific atoms. Your specific subatomic particles are the "brains" behind your three specific atomic particles. Your specific super subatomic particles are the "brains" behind your specific subatomic particles. Your specific super super subatomic particles are the "brains" behind your specific super subatomic particles, and so on, and so on, and so on....................

In the cosmos, smaller and faster encapsulations of matter are **"Always"**the "brains" behind larger and slower encapsulations of matter.

The Memory Of Matter

Every atomic particle "remembers" what to do in every specific, continuously balancing amount, and every specific, continuously balancing sequence of our universe's specific, continuously balancing amount, and specific, continuously balancing sequence of specific, continuously balancing environments, including specific, continuously balancing amounts of heat, and specific, continuously balancing amounts of pressure. The three specific atomic particles of biological matter did not "learn" how to become one of the specific atoms of biological life. They were not "taught" how to become one of the specific atoms of biological life. They "remembered" what to do when they were intentionally exposed to a specific amount, and a specific sequence of specific, continuously balancing environments, including specific, continuously balancing amounts of heat, and specific, continuously balancing amounts of pressure. Each specific organic molecule of biological life is produced identically. Each specific organic molecule of hydrogen rich methane, ammonia, and water vapor does not "learn" how, and is not "taught" how to become a mini protein, or mini nucleic acid that knows how to reproduce itself. They "remember" what to do when they are intentionally exposed to a specific, continuously balancing environment, including the specific, continuously balancing amount of heat, and the specific, continuously balancing amount of pressure of a specific "Bolt" of lightning. In "this" specific environment, specific organic molecules become attractive, or neutral, or repulsive to other specific organic molecules. In "this" specific environment, specific organic molecules are specifically attracted to other specific organic molecules, and are specifically neutral to other specific organic molecules, and are specifically repulsive to other specific organic molecules.

In "this" environment they simply "remember" what to do. Which is "accumulate", "interact", and "stay away".

Matter itself is a "memory" retention system. Every atomic particle's behavior, and every atom's behavior, and every organic, and every chemical molecule's behavior, is "continuous". This continuous behavior is based on "memory". All of these specific atomic accumulations "remember" what to do in "Every" specific environment of our universe.

The atomic neutron is the apex of the subatomic universe's evolution. It is the apex of cosmic memory retention. Our universe is continuously producing the post atomic neutron. It is the apex of our universe's evolution. Our universe, like every universe, produces the apex of cosmic memory retention by continuously producing a continuously accumulating specific, continuously balancing amount, and specific, continuously balancing sequence of specific memories. This is DNA

Everything your body does is based on memory. The continuous beating of your heart, the continuous cleaning of your blood, the continuous moisture produced to continuously protect your eyes, is all produced by the "memory" of matter.

All of your body's "memories" are stored in many different specific memory systems. Each specific memory system is in continuous communication with the each and every other different specific memory system. All of your body's "memories" are based on "experience". If one hundred generations of your family ate raw whole food, and had to carry this food up and down very steep mountains, "you" have strong legs, "you" have good balance, and "you" have strong teeth. This is all based on experience, and memory. If you're walking, and slip on a piece of ice, your body instantaneously begins adjusting and readjusting many complex balancing dynamics, while simultaneously introducing the precise amount of adrenaline to anesthetize your fall. This is all based on experience, and memory. When you fall asleep.................... and wake up hours later, everything your body did while you were asleep is based on experience, and memory. All of these memories are stored in specific memory warehouses, and "recalled" in specific ways. Just like if you smell something you haven't smelled in years, or you hear a song you haven't heard in years. This "trigger" activates specific memories of specific things that up until then, seemed completely forgotten. And yet, there it is. "All" of your "instincts" are produced by specific "triggers", and specific "memories".

"Memories" from other humans can be recalled if you inhale, absorb, or you incorporate specific accumulations of another human's un-reprocessed molecular compilations. If you have had an organ transplant, and afterword you suddenly prefer different food, or music, or sports, or television programs, or remember different things, or people, it's because your body's specific information translation system can simply translate your new organs specific accumulated memories of it's previous human. All of our "organs" are atomic cosmic "whole" components. They are also "brains" themselves, and are "**continuously**" interactive with the continuous flow of information entering and exiting our bodies. A heart does not watch TV. A liver does not listen to music. An "organ"does not eat pizza. But every one of our "organs" is continuously "eavesdropping" on what the eyes are doing, and what the taste buds are doing, and what the eardrums are doing through a specific amount, and a specific sequence of translators. Human "organs" remember differently than our grey brain, or DNA. But they "do" remember.

You may think you are someone from the distant past. That's because at one point in your life, you absorbed, or inhaled, or incorporated a specific amount of that person's specific un-reprocessed memory retaining molecular accumulations. At one point in your life, "molecularly", you "were" that person. Not "all" of them of course, but just enough to translate a perceived memory. "Every" dream you have ever had, includes a specific amount of someone elses un-reprocessed translated molecular memories. You have "never" seen a ghost. You just simply inhaled, and molecularly, you "remembered" that person. You simply translated their specific memory retaining un-reprocessed molecular compilations.

You may think you are someone from our Earth's future. This is impossible, because it's "always" "**now**", and it's "always" "today" on our Earth, and every other Earth. Some of our universe's earths are very young, like our Earth used to be. Some are the same. Some have advanced to the peak of human evolution, and countless others have been cosmicly reprocessed. In space, there are "many" specific memory retaining un-reprocessed molecular compilations from "many" advanced humans. Some of these specific memory retaining un-reprocessed

33

molecular compilations continuously sneak past our atmosphere, and continuously interact with our biosphere. If you have inhaled, or absorbed, or incorporated these specific molecules, and translated their specific memories, then you have seen the future. It is not the future of our particular Earth. It is the particular past of a much more advanced Earth than ours. But of course, all Earths evolve to this advanced state. Our Earth is no different. Every Earth's future is the same.

Biological Evolution

One of matter's constant patterns is to: "make the same", "make a specific amount", and "keep making them". Every "thing" that matter produces is produced "continuously", and is exactly the same as something else in the universe. There is a very specific, continuously balancing amount of each and every atomic cosmic whole replicant. Every "Earth" is a replicant. A specific, continuously balancing amount of Earths are in the identical evolutionary phase as our Earth, and a specific, continuously balancing amount of Earths are in every specific evolutionary phase of every Earth's specific evolutionary sequence.

Every replicant earth "continuously" produces a specific, continuously balancing amount, and a specific, continuously balancing sequence of specific biological "Replicants". There are no variables of atomic matter, or biological matter. There is "only" a tremendously large but specific "amount", and "sequence" of expected results. The different replicants on every different Earth may look a little different, but their atomic and molecular structures are "Identical". Each specific, continuously balancing amount, and specific, continuously balancing sequence of specific biological replicants produce a specifically different function for our Earth. Each and every one of our Earth's biological replicants are produced for the exact same reason. "Every" Earth's biologicle replicants are also produced for the exact same reason. "Every" replicant in our galaxy, and in our universe is also produced for the exact same reason. Every replicant smaller and larger "universe" is also produced for this exact same specific, identical reason. "Everything" is continuously produced, to continuously reproduce "Everything".

As biological matter accumulates, it becomes larger, and it also slows down to reveal what it's smaller behaviors are. "Every" larger accumulation of biological matter is simply a slow motion, or suspended animation transformed graph of "Every" smaller and faster biological accumulation.

Biological matter "survives" by outwitting and overpowering it's environment, to become the environment.

Each biological replicant reproduces by using two specifically different base components that produce a replicant of it's originators. Some biological replicants are androgynous, just like the universe. Each one is one whole replicant, that contains both of the "two" replicant reproducing components. Just like every earth.

Every biological organism produces an "up until this point" replicant of themselves by using "memory". This is DNA. DNA is your "up until this point" memory of everything. Your DNA remembers every shrimp you've ever eaten. Your DNA remembers everything about what "Every" shrimp you have eaten, has eaten. Your DNA remembers "Every" breath you're great, great, great, great, great, great, great grandmother took. Your DNA remembers "Everything" about the temperature and the pressure of the ocean in every generation of your three billion-year-old aquatic plant ancestors.

Your DNA remembers everything. There is a peak evolutionary, continuously balancing amount of DNA memories, and a peak evolutionary, continuously balancing sequence of DNA memories. This is the evolution for our earth's humans. This is the evolution for every replicant earth's replicant human like creatures. Every strand of DNA is produced exactly like a gold atom, and an organic molecule. In a specific, continuously balancing amount, and a specific, continuously balancing sequence of specific, continuously balancing environments. Every specific, continuously balancing environment includes a specific, continuously balancing amount of heat, and a specific, continuously balancing amount of pressure.

Emotion

Many atomic accumulations, or elements look like they are "still", and not moving. They may look still, but inside each and every atomic accumulation, there is "motion". Every specific electron in every single atom exhibits a "continuous" behavior. Each specific electron is continuously, externally encouraged, or stimulated to behave in a specific way, and each specific electron is continuously, internally motivated to respond in a specific way.

Every proton, and electron, and neutron's behavior is "constant want". When specific accumulations of atomic matter request other specific accumulations of atomic matter to respond, there may be no response at all. When the request is what a specific accumulation of atomic matter is "expecting", there is a "response". There is an "emotion". There is a continuous "want" of this specific request.

Complex accumulations of organic matter are simply specific transformations of continuous "want".

Every insect is "stimulated" by the sun. Some are internally motivated to move toward the sun, or away from it. They have to "want" to respond. A biological organism, as well as an electron, must continuously "respond" to survive. If a proton, as well as an ant, does not continuously respond to it's continuous external stimulation, it cannot survive. If a neutron, as well as an ant, does not continuously respond to it's continuous specific internal "wants", it will not survive.

Every atomic, and biological accumulation of matter is continuously, externally encouraged, "**And**" continuously, internally motivated.

You cannot control if you are thirsty or not. You can choose not to drink, but internally, you will be continuously reminded. This is a specific transformation of continuous "want". You cannot control if you're hungry or not. You can choose not to eat, but internally, you will be continuously reminded. This is a specific transformation of continuous "want". You cannot choose which cells your body continuously produces. This is a specific transformation of continuous "want". You cannot control if you're sleepy or not, you can choose not to sleep, but internally, you will be continuously reminded. This is a specific transformation of continuous "want". When you have to go to the bathroom, you can choose not to go, but internally, you'll be continuously reminded. This is a specific transformation of continuous "want".

You cannot control what will spontaneously make you burst out into laughter. This is a specific transformation of continuous "want". Every sentiment you feel, every longing, every wish, every desire, every craving, every passion, every single emotion you have ever felt, is simply a specific transformation of continuous "want".

If a stray cat sees you, he quickly runs away because he is afraid. If a wild squirrel sees you, he runs up a tree because he is afraid. If you see a snake in the grass, he slithers away quickly because he is afraid. If you brush up against a large black ant once or twice, he will run away because he is afraid.

The ant's continuous behaviors that produce his continuous running are specific transformations of continuous "want". The ant's continuous fear is a specific transformation of continuous "want". This ant is continuously afraid, so he continuously runs. He continuously runs, because he continuously "wants" to survive. This ant may also continuously attack your finger. This is a specific transformation of continuous "want". Every biological organism is continuously internally motivated to survive, to continue. Continuing is a specific transformation of continuous "want". Biological organisms, and atomic accumulations continuously "want" to continuously "want". If you turn your houseplant around in your window at home, the leaves will turn back toward the sun. This is a specific transformation of continuous "want". Every continuously shining star is a specific transformation of continuous "want". Every specific transformation of continuous "want" in our universe, is produced by continuous repetitive cycles, of specific amounts, and specific sequences of continuous;

"Attraction", (+)----------(**N**)----------(-) and **"Repulsion"**.

Reproduction

Everything in our universe is a "reproduction". Every galaxy, every star, every earth, every human, every molecule, every atom, and every atomic particle. Atomic reproduction is produced by two accumulations of atomic matter "engaging" to produce a very small atomic matter accumulation regulating "seed", or "egg". For a human, the two "engaging" accumulations of matter are the sperm, and the egg. For an oak tree, it is the stamenate, and the pistilate. For our universe, and every other smaller and larger universe, the two reproductive engaging accumulations of atomic matter are the "Human", and the "Diamond". The resulting atomic matter accumulation regulating seed reproduces a replicant of the originator by continuously circulating smaller atomic matter in and out of it's overall system, and continuously accumulating the intended atomic matter. Matter produces a specific amount of transformations of this dynamic. Many atomic reproductive patterns are very different, and some atomic reproductive patterns are very similar.

"Every" universe's reproductive pattern is very similar to the oak tree's reproductive pattern. Every universe's replicants, as well as every universe itself, reproduce themselves by using specific accumulations of smaller matter.

"One" oak tree begins producing it's "many" smaller replicant acorns "inside" of the tree, using the stamenate, and the pistilate.

"One" universe begins producing it's "many" smaller replicant neutrons "inside" of the universe, using the human, and the diamond.

"Every" replicant acorn works it's way to the very outskirts of the tree.

"Every" replicant neutron works it's way to the very outskirts of the universe.

"Every" replicant acorn has an "up until this point" warehouse of replicating information.

"Every" replicant neutron has an "up until this point" warehouse of replicating information.

"Every" replicant acorn is very puny compared to the tree that produced it, and looks "nothing" like it's producer.

"Every" replicant neutron is very puny compared to the universe that produced it, and looks "nothing" like it's producer.

After the reproduction cycle is complete, "Every" replicant acorn is "released". After about one hundred cycles of continuous specific smaller matter accumulation, it produces a replicant of it's producer.

After the reproduction cycle is complete, "Every" replicant neutron is "released". After a specific amount of cycles of continuous specific smaller matter accumulation, it produces a replicant of it's producer.

"Every" acorn represents the oak tree's present level of cosmic maximum complexity retention. "Every" acorn represents the oak tree's present cosmic "direct energy output %, to indirect energy output %, in relation to overall energy input", continuously balancing equation. Every acorn represents the oak tree's continuously balancing equation of cosmic "Health", "Strength", and "Wisdom". Every acorn represents the oak tree's present level of cosmic evolution.

"Every" neutron is the cosmic representation of maximum complexity retention. "Every" neutron is the cosmic maximum "direct energy output %, to indirect energy output %, in relation to overall energy input", continuously balancing equation. "Every" neutron is the continuously balancing equation of maximum cosmic "Health", "Strength", and "Wisdom". "Every" neutron is the apex of cosmic evolution.

There are many different types of oak trees. Each type of oak tree is the balancing point representation of their individual environmental experiences over millions of years. Each time the environment changes, each specific replicant changes. When the environment stabilizes, so does the balancing point representation of the replicant.

Every complex biological organism produces a smaller replicant "seed". When two frog reproductive components "engage", they can produce hundreds of replicants. These replicants are very small, and look nothing like their generator, yet they "all" have an information warehouse that enables them to replicate their generator. When two butterfly reproductive components "engage", they can produce hundreds of replicants. These replicants are very small, and look nothing like their generator, yet they all have an information warehouse that enables them to replicate their generator.

Atomic matter "encourages" itself to reproduce in a specific, continuously balancing environment. Each specific, continuously balancing environment includes a specific, continuously balancing amount of heat and pressure. This specific environment continuously encourages the two reproductive components to accumulate into two specific components that "want" to "engage". These two specific components "want" to come "together". This specific dynamic is "attraction". This specific "attraction" is produced in a specific, continuously balancing environment of a specific, continuously balancing amount, and sequence of continuous "repulsion".

Atomic matter has produced a way to "Attract" a specific amount of space hydrogen together to become a specific star. Matter has produced a way for a turtle to be "attracted" to another turtle. Matter has produced a way for a gold atom to be "attracted" to another gold atom.

The reproduction of a star, and a snake is produced the same way. The reproduction of a gold atom, and a tree is produced the same way.

To understand how a star comes together, or how atoms come together, we simply have to translate why biological creatures come together. The behavior of biological matter is simply a larger and slower transformed motion result graph of "smaller" atomic matter's behavior, and a smaller and faster transformed motion result graph of "larger" atomic matter's behavior. To understand the intent of matter, you simply have to translate it's many accumulation's behaviors. To understand "how" atomic matter reproduces itself, you simply have to translate one of it's biological replicant's continuous reproductive behaviors. Even if we could communicate with a star, or a plant, or a spider, they still couldn't tell us how they reproduce. To translate the continuous reproductive behavior of biological matter, is to translate the continuous reproductive language of atomic matter, and matter itself, and it's "attractive" formula of reproduction.

Biological Reproductive Communication

The reproductive behavior of "all" atomic matter is the same. Atomic matter simply speeds up and transforms when it is in small accumulations, and it slows down and transforms when it is in larger accumulations. The reproductive communication of all complex biological organisms is produced by many different smaller micro-organisms existing as one "whole".

All complex biological organisms reproductively communicate. As these complex biological organisms accumulate, they become even more complex, and so do their reproductive communication dynamics. Smaller less complex biological organisms reproductively communicate with magnetism, and electric. As they accumulate and become more complex, they produce an even more complex reproductive communication with Touch, Taste, Smell, Sound, and Sight. Plants are just as complex as animals, if not more complex than animals. But it is impossible to translate their reproductive communication. We cannot smell what a tree smells. We cannot hear what a tree hears. We cannot see what the tree sees. Insects are just as complex as animals and plants, if not more complex. But it's impossible to translate their reproductive communication. We cannot smell what an insect smells. We cannot hear what an insect hears. We cannot see what an insect sees, but we can easily translate an animals reproductive communication, because that's exactly what we are.

The "mastery" of an animals environment is "always" on display. Each individual animal represents a certain level of "Health", "Strength", and "Wisdom". For an animal, if you can continuously acquire enough food to survive, and you continuously keep from being food for something else, you continuously represent a certain level of Health, Strength, and Wisdom. The acceptable continuous observation and experience of this dynamic is "Stimulating". It is "Attractive".

Atomic matter "always" uses attraction to bring two cosmic whole components together. Atomic matter "always" uses repulsion to encourage the attraction. When we observe the reproductive behavior of complex biological animals, we are simply looking at the larger, slower, and transformed identical formula that brings atomic particles "together", and also atoms, and molecules, and organic molecules, and complex organic molecular compilations, and every other less complex smaller biological organism "together". We are also looking at the smaller and faster transformed identical formula that brings a "planet" together, or a "star" together, or a "galaxy" together. We simply have to translate it.

All of the clues to atomic matter's reproductive formula are revealed by "Birds". Birds do not know why or how they reproduce. Birds simply remember an observed formula that achieves a continuous "want", and they repeat it. Birds evolved from dinosaurs, and they have been reproductively communicating for millions and millions of years. Over these millions and millions of years, their reproductive formula becomes very diverse. Some birds exhibit almost no visible reproductive communication at all, and some birds, unintentionally of course, reveal the "Entire" reproductive formula of atomic matter, and matter itself. The "continuous" reproductive communications of "Birds" are the "Continuous" clues.

For the most complex birds, it is the female that "chooses" it's reproductive counterpart. At a specific point, the "Adult" male bird becomes "Continuously" reproductively attracted to the female. The male bird must become "irresistible" to the female. She cannot "think". She must "respond". She must be overwhelmingly compelled to "Choose". "**She**" must be "Attracted". "**He**" is continuously "attracted", like every other adult male biological organism.

For a male bird to survive every day, he must continuously outwit, and overpower his environment. The success rate of this dynamic is continuously on display. It is "him". Every "more" complex male bird uses all five external senses to reproductively communicate. Some male birds use more of one sensory stimulant than another. The most complex male birds use as much of "all" of them as they can.

"Every" different "type" of adult male bird becomes the complete transformed graph of atomic matter's reproductive communication formula. Each different "type" of bird speaks a different language. Some birds do not talk much at all. Some birds are simply white, or black, and some birds barely look at each other. These birds exhibit too small of a graph of reproductive communication to translate it properly. There are other birds that simply "are" the complete transformed graph of atomic matter's reproductive communication formula.

Some male birds are not very visually stimulating, or "colorful". These types of birds rely on their song, or voice to attract a reproductive counterpart. There is a specific type of less colorful bird that will perch at the top of an antenna, or telephone pole to reproductively communicate his intention. This type of bird will "sing". He sings so many songs, it's hard to tell if he even has a song of his own. He sings as many "other birds" songs as he can remember, and he sings them in a "row". In a "sequence". Each specifically different bird song, in his complete bird song sequence, is a specific amount, and a specific sequence of specific sounds. The specific bird that sings each specifically different song, sings his own reproductively attractive song in a continuous, repetitive cycle. He sings a specific amount, and a specific sequence of specific sounds over, and over, and over, and over again. The other bird uses this specific amount and sequence of specific sounds once, or twice, and sometimes a few times in his specific amount, and specific sequence cycle of specific bird songs. This new complete song is now a continuous, repetitive cycle, of a specific, continuously balancing amount, and a specific, continuously balancing sequence of specific, continuously balancing amounts, and specific, continuously balancing sequences, of specific sounds, which produces the "Attractiveness".

There are many male birds that reproductively communicate by using many different colors, "and" sounds. There are many male birds that reproductively communicate by using "dancing", as well as colors and sounds. There are many male birds that reproductively communicate by using smells and tastes, as well as colors and dancing and sounds. It's hard to tell just how much smell and taste is used, if you're not a bird.

There are only a few male birds that build exhibition structures for their reproductive communication, as well as using colors, sounds, smells, tastes, and dancing. There is one specific male bird that builds an elaborate stage around a tree or tree stump. He builds this stage with sticks and twigs placed around the base of the tree. He also enhances the stage with carefully placed mosses and berries. He will continuously adjust and readjust the mosses and berries until it is just right. Then he will begin to continuously sing, as he continuously dances around and around the stage.

Every feather of this bird, every color of this bird, every smell of this bird, every sound it makes, every dance step it takes, every stick it picks up for the stage that it makes, represents his present levels of "Health", "Strength", and "Wisdom". The "wisdom" to continuously eat the right kind of food. The "wisdom" to continuously eat the right amount, of the right kind of food. The "wisdom" to continuously not become food for something else. The "wisdom" to select and collect the right sticks and twigs for the stage. The "wisdom" to select and collect the right berries and mosses for the stage. The "wisdom" to remember, and repeat previously experienced reproductive dances and songs. The "strength" to continuously protect itself, or continuously run away from every predator. The "strength" to lift and carry the many sticks and twigs to the staging area. The "strength" to arrange the many sticks and twigs into a stage. The "strength" of the stage itself. The "strength" of his dance steps. The "strength", or the "intensity" of his song. The "health" of his feathers. The "health" of his feathers colors. The "health" of his physical self. The "health", or longevity of his dance. The "health", or longevity of his song, and even the "health" of the decorative mosses and berries is his "continuous" reproductive communication. The female bird continuously observes this exhibition, and she is "attracted", or not. If "any" one of these components is not in the right amount, or is not in the right sequence, or is not as "attractive" as "all" of the other components, his chances of reproduction are less likely. To be reproductively successful he must continuously exhibit an attractive taste, an attractive smell, and attractive feathers, and attractive colors, and an attractive shape, and an attractive size, and an attractive stage, with attractive mosses, and attractive berries, and an attractive dance, and an attractive song. This is "irresistible" to the female bird.

The "irresistibility" is in the "balance".

"All" the right components, in all the right amounts, in just the right sequence, is "irresistible" to the female bird. This identical transformed dynamic is how the adult female bird becomes "irresistible" to the adult male bird. At the peak of this dynamic, each bird's continuous assessment of each other erases all of their previous thoughts, and replaces it with "one" continuous "thought". "One" continuous "want". Reproductive communication. It's the same as when you and a few of your friends are talking, and continuously "thinking" about a certain subject. As you are talking, and continuously thinking about this specific subject, a beautiful man or woman "walks" past your group. In one split second, each one of your thoughts are completely erased, and replaced with "one" continuous thought. "One" continuous "want". Reproductive communication. To come together. To "Accumulate".

The bird and human reveal that for biological matter, as well as matter itself to be "continuously" attracted to another accumulation of biological matter, or matter, it must accumulate into this irresistible state. The bird and the human reveal that one specific accumulation of biological matter, as well as matter itself "chooses" one specifically different accumulation of biological matter, or matter, to "come together" and produce a replicating "seed". This bird also reveals that atomic matter, as well as matter itself becomes "irresistible" to itself by using a specific, continuously balancing amount, and a specific, continuously balancing sequence cycle, of specific, continuously balancing amounts, and specific continuously balancing sequence cycles, within a specific, continuously balancing amount, and a specific, continuously balancing sequence cycle of specific, continuously balancing environments, including specific, continuously balancing amounts of heat, and specific, continuously balancing amounts of pressure.

Each year is a specific, continuously balancing, repetitive amount, and sequence cycle, of specific, continuously balancing amounts of heat, and specific, continuously balancing amounts of pressure. Each day is a specific, continuously balancing, repetitive amount, and sequence cycle, of specific, continuously balancing amounts of heat, and specific, continuously balancing amounts of pressure. When this bird is a baby and in the nest, he continuously "accumulates". He starts his accumulation at a specific point during the earth's specific, continuously balancing, repetitive amount, and sequence cycle of days. He begins to "accumulate" at a specific point during each day's specific, continuously balancing, repetitive amount sequence cycle of heat and pressure. Each day of his young accumulation is produced by these specific, continuously balancing, repetitive amount and sequence cycles, and the specific, continuously balancing amount, and the specific, continuously balancing sequence cycle of food his parents bring him, and his specific, continuously balancing internal amounts and sequence cycles of heat and pressure. As he "accumulates" into a young adult, he will encourage his "accumulation" by specifically regulating his own continuous intake of a specific, continuously balancing amount, and a specific, continuously balancing sequence cycle of food. After he has accumulated into an adult, he is attracted to the female. He must now be Healthy enough, Strong enough, and Wise enough to himself, "become" his own "irresistible" accumulation, of "one" specific, continuously balancing repetitive amount and sequence cycle, of specific, continuously balancing amounts, and specific, continuously balancing sequences, of specific components. These specific components are his singing and dancing. He will repeat his specific song, and his specific dance, over, and over, and over again. If all the right components, are in all the right amounts, in just the right sequence, he will be "irresistible". If his complete, continuously balancing, repetitive, reproductive communication amount and sequence cycle is not successful, he will repeat the cycle until he is successful, or too tired to repeat it.

This "bird" and the "human" reveal that biological matter, as well as matter itself, is "attracted" to a continuously balancing retention, or equation of maximum Health, Strength, and Wisdom. The more "balanced" the atomic, or biological compilation is, the more "attractive" it is. This "bird" and the "human" reveal that biological matter, as well as matter itself, is "continuously" assessing a reproductive counterpart. This "bird" and the "human" reveal that biological matter, as well as matter itself, is continuously "calculating" to produce a continuously balancing equation result. When the calculation is "right", irresistible attraction will occur.

"Calculation" is the language of matter.
"Sound" is the calculation of matter.
"Balance" is the intent of matter.

Matter reveals that within a specific, continuously balancing environment including a specific, continuously balancing amount of heat, and a specific, continuously balancing amount of pressure, two specifically different, continuously balancing amounts, and continuously balancing sequences of specific, continuously balancing repetitive sound cycles, produce the "Attraction" of matter, and two specifically different, continuously imbalancing amounts, and continuously imbalancing sequences of specific, continuously balancing repetitive sound cycles, produce the "Repulsion" of matter. "Repulsion" is how atomic matter disintegrates. "Attraction" is how atomic matter accumulates. This is how atomic matter reproduces itself.

The reproduction of atomic matter is produced with "Attraction". In a specific, continuously balancing environment including a specific, continuously balancing amount of heat, and a specific, continuously balancing amount of pressure, a specific, continuously balancing amount, and a specific, continuously balancing sequence cycle of "Health", "Strength", and "Wisdom" is irresistible to atomic matter. This is "why" all atomic matter accumulates.

"Reproduction" is simply the continuous result of matter's continuously accumulating behavior.

All of the "patterns" of matter are simply transformed motion results of smaller and faster, or larger and slower accumulations of matter. All of the "patterns" of accumulated atomic matter are simply "tools" that matter itself uses to continuously reproduce itself. Atomic matter continuously makes these tools that continuously reproduce itself out of "one" specific medium. Itself.

These are the "tools" of matter.

These are the constant "patterns" of matter.

The Coil

The "coil" is matter's balancing "tool".
It is also the prime motion of matter.
When you see a maple tree seed falling to
the ground, you are watching the prime
motion of matter. A maple tree seed falls
like a helicopter, spinning to the ground.
The entire flight of this seed is a "coil". You
don't see the coil because the seed doesn't
leave a trail. But if you make a solid line
graph of it's flight, it is a coil. When a
squirrel jumps from one tree to the other,
his tail "spins" a little to stabilize his flight.
If you were to make a solid line graph of his
tails behavior during this flight, it would be
a coil. A balancing, coiling motion. There
is a specific amount of transformations of
the coil in in our universe. They all may
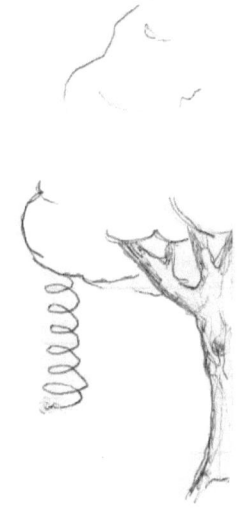
look a little different, and some may move much faster than others,
and some may be in suspended animation, but they "all" provide the
same function for matter. The coil helps to balance the cosmic "whole"
component that it is presently a part of, and it also helps to balance
the larger encapsulating cosmic whole component, that it's present
whole component, is a micro component of.

The double helix of DNA is two coils. There are many snail shells that are coils. The whirlpool of water that goes down your sink is a coil. There are many different spiderwebs that are a coil. A spiral galaxy like ours is two or more coiling arms.

The behavior of the planets, and moons, and stars, and galaxies, is the behavior of our subatomic particles, only much, much, much slower. Our moon continuously orbits our earth. Our earth continuously orbits our sun. Our sun continuously orbits our galcleous, and our galcleous continuously orbits the ucleous. If you were to make a solid line graph of the moon's orbit around the earth during the earth's orbit around the sun, you would "see" the moon's orbit as a coil wrapped around the earth's orbit. You would also "see" the earth's orbit as a coiling circle.

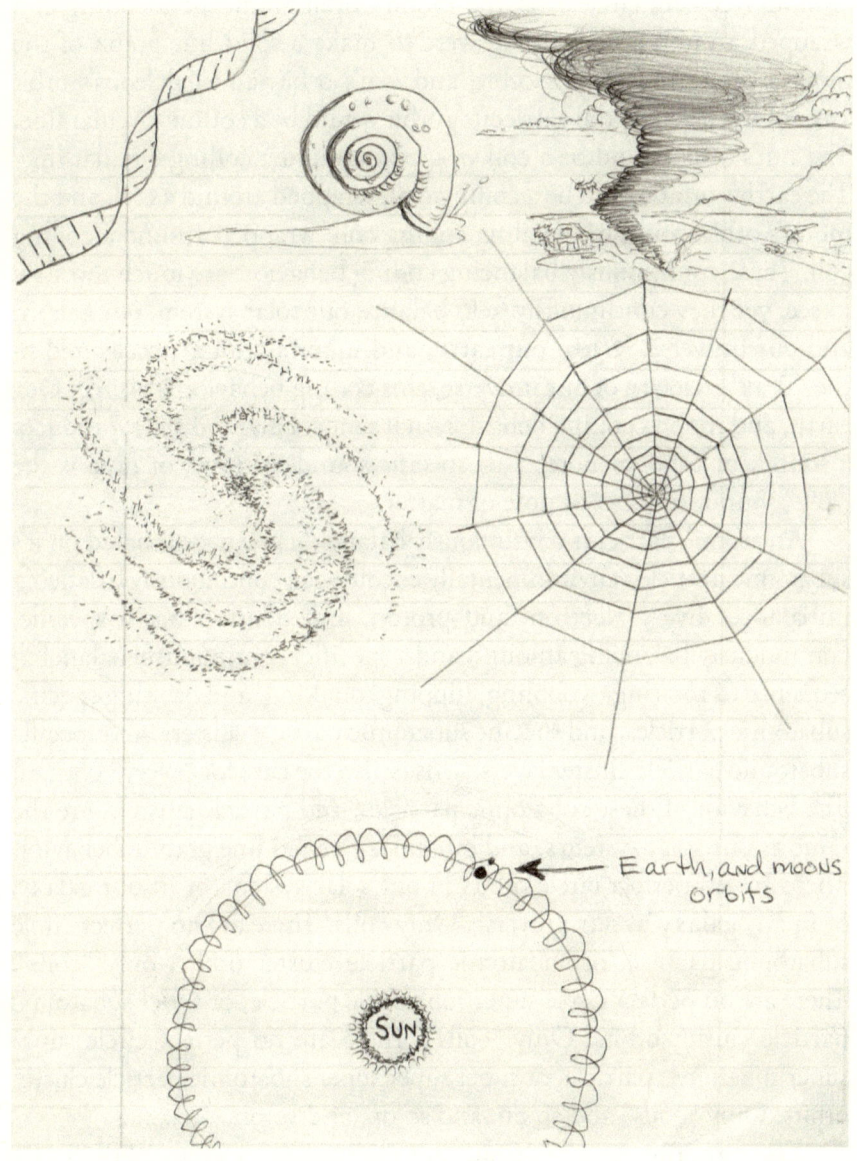

Earth, and moons orbits

SUN

If you were to make a solid line graph of the moon's orbit, and earth's orbit, and the sun's orbit around the galcleous, the sun's orbit would be a coiling circle. The earth's orbit would be a coil wrapped around a coiling circle, and the moon's orbit would be a coiling coil wrapped around a coil. If you were to make a solid line graph of the moon's orbit, and earth's orbit, and sun's orbit, and galcleous's orbit around the ucleous, the galcleous's orbit would be a coiling circular line. The sun's orbit would be a coil wrapped around a coiling circular line. The earth's orbit would be a coiling coil, wrapped around a coil, and the moon's orbit would be a coiling coiling coil, wrapped around a coiling coil. These continuously balancing coiling behaviors are much too slow to see, yet they continuously help balance our solar system, our galaxy, and our universe. When our earth, and moon's orbit is accelerated to the "Peak" velocity of our universe, this coiling behavior is visible. Our earth, and moon's coiling behavior, in it's most balanced state, produces a sound, or tone, or note. This specific sound, or tone, or note is the "key" sound, or tone, or note of matter.

An atomic particle's continuously balancing behavior is based on it's subatomic particle's environmentally encouraged continuously balancing imbalance. Every electron, and proton, and neutron has a specific, continuously balancing amount, and a specific, continuously balancing sequence of rotating, wobbling, flipping, quaking, and orbiting specific subatomic particles, and specific subatomic particle clusters, and specific subatomic particle cluster layers. This is also the case for "every" photon. The behavior of these subatomic particles, and particle clusters, are the same as our solar system's, and our galaxy's solid line graph's behavior. There are no perfect circle orbits in our solar system, or in our galaxy, or in any galaxy in our universe. Only coils. There are no perfect circle subatomic particle, or subatomic particle cluster orbits, only "coils". There are no perfect circle super subatomic particle, or super subatomic particle cluster orbits. Only "coils". There are no perfect circle super super subatomic particle, or super super super subatomic particle cluster orbits. Only "coils" and so on, and so on, and so on....................

Every proton, and neutron, and electron, and every photon is a specific, continuously balancing unified amount, and a specific, continuously balancing unified sequence of specific unified subatomic layers, of specific unified subatomic particle clusters, of specific unified subatomic particles. They are only stable in their particular unified state. When science separates the atomic particle, the "**coil**" is always represented.

Every specific, continuously balancing frequency of the electromagnetic spectrum is two propagating "coils" of specific atomic activity. Our sun continuously emits "every" propagating coiled specific, continuously balancing frequency of electromagnetic radiation. This specific, continuously balancing coiled information continuously helps balance our Earth. Our Earth, in turn, like every other advanced earth, will continuously help balance our galaxy with humans, and diamonds. Our galaxy, in turn, will continuously help balance our universe.

Atomic matter "continuously" reproduces itself, by continuously balancing itself, with continuous, repetitive "patterns", or "tools" of itself, produced by accumulation transformations of it's smaller self.

Out Half Around Rest.
Half Around Back Rest.

This continuously balancing motion is the "**prime**" motion of matter. It is the continuously balancing motion of every coiling orbit of accumulated matter. There is a specific, continuously balancing amount of different, specific, continuously balancing "motions" in our universe. They are all produced by a specific, continuously balancing amount, and a specific, continuously balancing sequence of specific, continuously balancing accumulation transformations of "**one**" specific, continuously balancing motion.

In our galaxy, there are "many" different, specific, continuously balancing motions. On our earth, there are "many" different, specific, continuously balancing motions. "All" of these different, specific, continuously balancing motions are simply specific accumulations of "many" different "smaller", specific, continuously balancing motions. The origin of "all" of our universe's different, specific, continuously balancing motions start from the same place. The ucleous. To determine the origin of the ucleous's specific, continuously balancing unifying motions, we have to start with the first independent behavior that is exhibited "from" the ucleous. The subatomic universe is the end of one universe, and the beginning of our universe. The subatomic universe is the end of one complete cosmic accumulation transformation sequence cycle of matter, and the beginning of another.

The ucleous continuously produces atomic neutrons. Atomic neutrons are the finished product of the subatomic universe. It is the subatomic universe's "seed", or "egg". Each and every atomic neutron comes

Out

of the subatomic universe. The subatomic universe continuously pushes "**out**" atomic neutrons. In our universe, and in every other smaller encapsulated universe, and in every other larger encapsulating universe, the first behavior of accumulated matter is "**out**". As the slowly rotating ucleous continuously pushes "**out**" atomic neutrons, they all continuously orbit

Around

the ucleous. As the atomic neutrons continuously accumulate, they become a unified layer of orbiting atomic neutrons. This is the Univator. The continuous production of atomic neutrons continuously increases the size of the univator. When the continuous pressure of our entire universe becomes too great, and the continuous attraction of the ucleous becomes too strong, the continuously accumulating univator's size will stop increasing, and begin decreasing. "In between" the univator's continuously increasing size, and it's continuously decreasing size, there is a

Rest.

It may be one trillionth of the second, or it may be one trillion years. Nevertheless, there is a "**rest**". This is the first of two "**rests**" in the journey of every univator neutron. Every atomic neutron in the now resting univator has completed "**half**" of it's first journey into our universe. Every atomic neutron that continuously comes "**out**" of the ucleous spends the first "**half**" of it's first journey into our universe orbiting "**around**" the ucleous. It's journey "**out**" from the ucleous is now at "**rest**".

Out half around rest.

As the mass of the univator continuously increases from the ucleous's continuous production of the atomic neutron, it begins to collapse on itself and the ucleous because of it's new unified masse's reaction to the specific, continuously balancing "Repulsion" of our universe, and the specific, continuously balancing "Attraction" of the ucleous. Each atomic neutron of the collapsing univator spends the second "**half**" of it's journey into our universe slowly orbiting "**around**" the ucleous, and returning "**back**" to the ucleous.

As the neutron filled univator continues to collapse on itself and the ucleous, it comes to a point where it's continuous collapsing stops. This is the second "**rest**" in the complete journey of every univator neutron. This second "**rest**" may last one trillionth of a second, or one trillion years, nevertheless, there is a "**rest**". After this rest, there is a tremendous explosion of atomic particles. This is the Univation. Or "The Big Bang".

Out half around rest. Half around back rest.

This is the continuous journey of every particle of matter. If an atomic particle is disassembled during the Univation's explosion "out", it simply turns "around" and returns "back" to the ucleous to be reprocessed into an atomic neutron. An atomic particle may become part of an assembled cosmic "whole" component that advances "out" and "around" two, or three, or maybe even one hundred larger encapsulating universes. No matter how far "out" and "around" it travels, if it's present cosmic whole component becomes disassembled, it simply returns "back" to it's original universe.

"Out half around rest, half around back rest", is the cosmic translation of Beginning, Middle, and End. The beginning of every cosmic cycle is "out". The middle of every cosmic cycle is "half around rest, half around", and the end of every cosmic cycle is "back rest". In our universe, and in every other smaller and larger universe, there is a specific amount of specific accumulation transformations of the journey of "out half around rest, half around back rest", which is the continuous journey of all matter.

Each individual, specifically different continuously balancing motion of rotating, and wobbling, and flipping, and quaking, and orbiting is "out half around rest, half around back rest".

The electric company sends a current "out" of the generator. It travels "half around" the circuit to your light bulb filament. This is the first "rest". Then the current travels "half around" the circuit "back" to the generator to "rest".

A plucked guitar string's behavior is "out half around rest, half around back rest".

A tapped tuning fork's behavior is "out half around rest, half around back rest".

Magnetism travels "out" of the magnet, then "half around" the magnetic field. This is the first "rest". Then it travels "half around" the field "back" to the magnet to "rest".

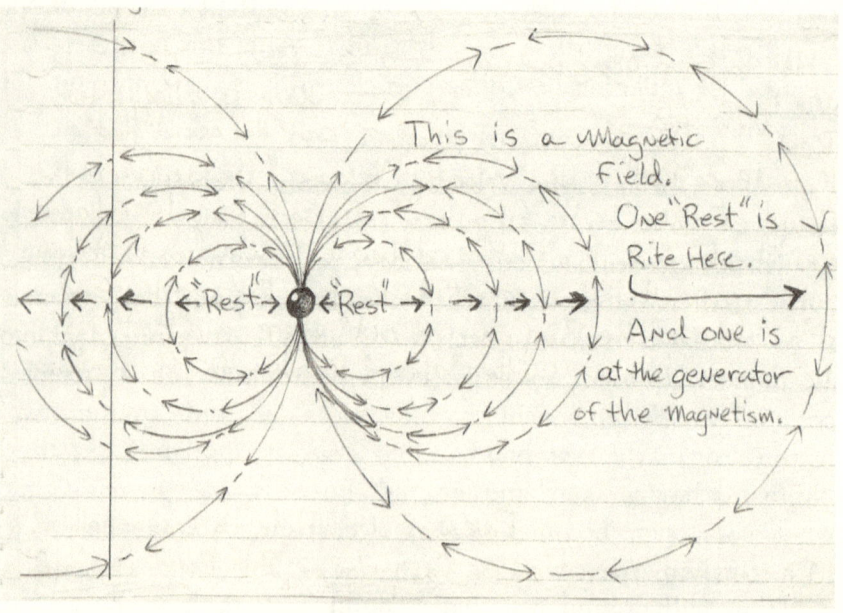

Our Earth travels "out" and "half around" to the furthest point of it's orbit. This is the first "rest". Then it travels "half around" it's orbit "back" to "rest" at the closest point of it's orbit around the sun.

Water vapor travels "out" of an ocean. Then it travels "half around" the earth. This is one of many "rests". Then it travels "half around" the Earth "back" to it's original ocean to "rest".

Lava travels "out" of the earth. Then it travels "half around" it's complete journey as a piece of the earth's crust. This is the first "rest".

Basically, once lava hardens it is at "rest". Then it travels "half around" it's complete journey "back" to the molten inside of the earth to "rest".

A bee travels "out" of it's hive. Then it travels "half around" it's environment to achieve it's goal. This is the first "rest". Then it travels half around it's environment "back" to the hive to "rest".

An octopus travels "out" of an enclosure. Then it travels "half around" it's environment to achieve it's goal. This is the first "rest". Then it travels "half around" it's environment "back" to an enclosure to "rest".

A bird travels "out" of it's nest. Then it travels "half around" it's environment to achieve it's goal. This is the first "rest". Then it travels "half around" it's environment "back" to it's nest to "rest".

An ant travels "out" of the ant colony. Then it travels "half around" it's environment to achieve his goal. This is the first "rest". Then it travels "half around" it's environment "back" to it's colony to "rest".

Your blood travels "out" of your heart. Then it travels "half around" your body to achieve it's goal. This is the first "rest". Then it travels "half around" your body "back" to your heart to "rest".

You travel "out" of your home in the morning. Then you travel "half around" your environment to achieve your goal. This is your first "rest". Then you travel "half around" your environment back to your home to "rest".

Every subatomic particle travels "out" and "half around" to the furthest point of it's orbit. This is the first "rest". Then it travels "half around" it's orbit "back" to "rest" at the closest point of it's orbit.

Every super subatomic particle travels "out" and "half around" to the furthest point of it's orbit. This is the first "rest". Then it travels "half around its orbit "back" to rest at the closest point of it's orbit, and so on, and so on, and so on....................

For subatomic, and super subatomic, and super super subatomic and so on... particles, the first "rest" is the exact point between acceleration, and deceleration. The second "rest" is the exact point between deceleration, and acceleration. This dynamic is the same for our Earth's orbit around the sun, and our sun's orbit around the galcleous, and the galcleouse's orbit around the ucleous.

When you stand in an open field, and you toss a ball "up" in the air, you are really not tossing it "up". Your tossing it "out" from the earth. In the cosmos, there is no such thing as "up", only "out". In the cosmos, there is no such thing as "down", only "back". In the cosmos, there is no such thing as "up" and "down", only "right side up", and "up side down". When you stand still on the equator, you are traveling about one thousand miles an hour with the earth. When you toss the ball "out" from the earth, it will travel "half around" then "rest". Then it will travel "half around" and "back" to your hand to "rest". This arcing, solid line graph of a tossed ball's complete journey is accentuated, but no matter where you toss a ball "out" from the earth, the solid line graph of it's complete journey is a hairpin "Arc".

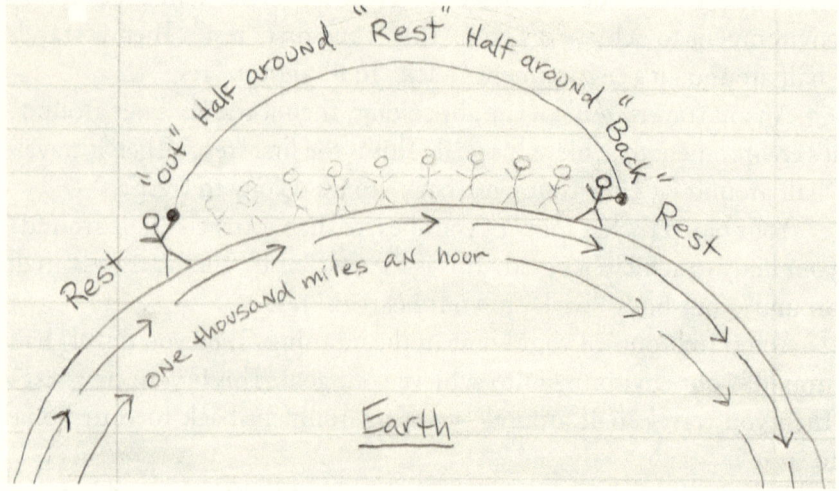

Even if you toss the ball only one inch into the air, the solid line graph of it's center of masses journey is a micro hairline arc. Even if you stood on the North, or South Pole and tossed the ball "out" from the earth, it still will travel "half around" then "rest", then it will travel "half around" and "back" to "rest". This is the behavior of a tossed ball on every planet, and every moon, and every star, and every galcleous. This is the behavior of a tossed ball "out" from the ucleous.

In the cosmos, there is no "first motion" of matter. There is no starting point of matter's motion. In each universe, there is only a specific, continuously balancing amount, and a specific, continuously balancing sequence, of specific, continuously balancing accumulation transformations of "out half around rest", "half around back rest".

The "Prime" motion of matter.

The Ramiform

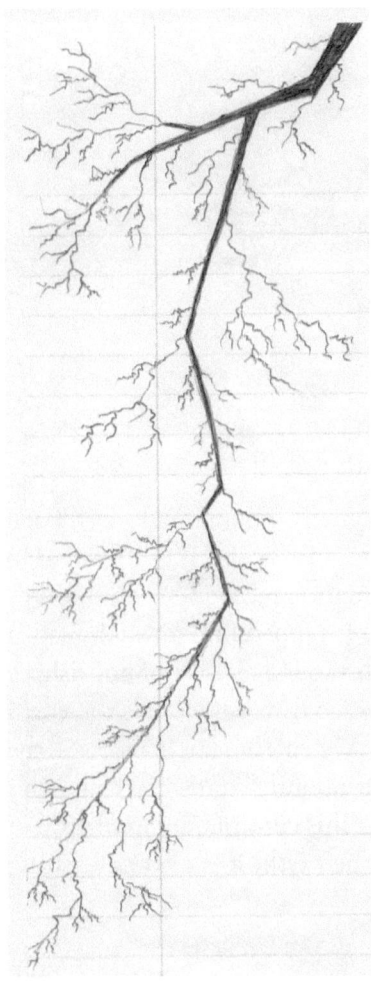

The "ramiform" is matter's information transportation "tool". There is a specific amount of specific transformations of the ramiform in our universe. They all may look a little different, and some may move much faster than others, and some may be in suspended animation, but they "all" provide the same function for matter. The ramiform transports nourishing information from a smaller encapsulated cosmic whole component, to a larger and encapsulating cosmic whole component. The ramiform also transports nourishing information from a larger encapsulating cosmic whole component, to a smaller encapsulated cosmic whole component. The ramiform can transport outgoing information "continuously". The ramiform can transport incoming information "continuously". The ramiform can transport incoming, and outgoing information "continuously". The ramiform can itself "be" the continuously incoming, and continuously outgoing information. The ramiform's base is "always" connected to an information transmitter. The ramiform's base is "always" connected to an information receiver. The ramiform's base is "always" connected to an information warehouse. The information warehouse, and receiver, and transmitter that produces the ramiform, is "always" at the center of the ramiformic activity.

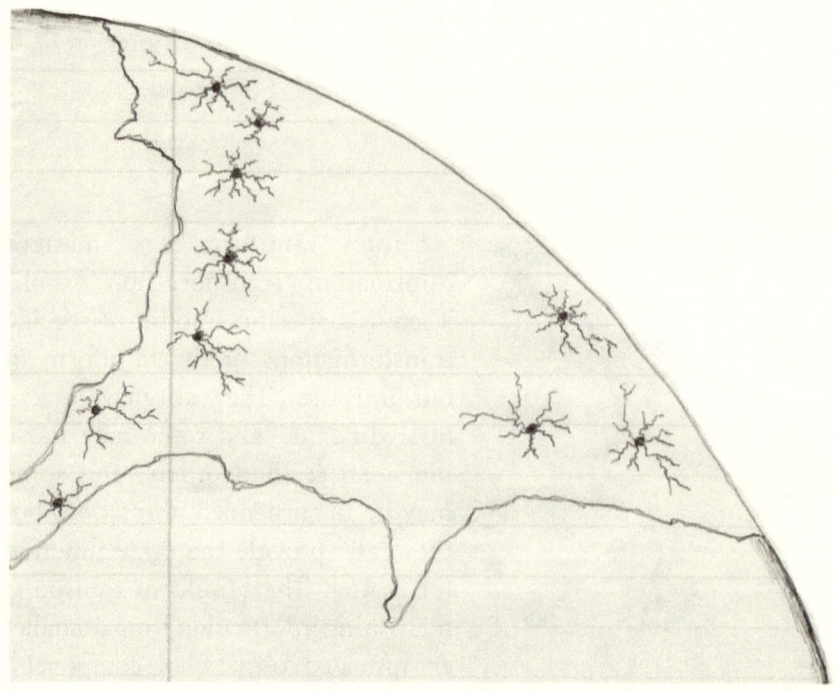

Every earth has two pole glaciers, and many mountain glaciers. Each one is a specific information warehouse, with a specific information transmitting system, and a specific information receiving system. Every earth has many volcanoes. Each volcano is tapped into a warehouse of the Earth's molten information. Each volcano is a receiver of nourishing information for the Earth. Each volcano "continuously" tells the earth what it's exterior heat and pressure is. When the earth receives the proper information, the volcano will "continuously" transmit the Earth's nourishing molten information response. This continuous circulation of information is not only nourishing for the earth, it is also nourishing for our galaxy, and for our universe.

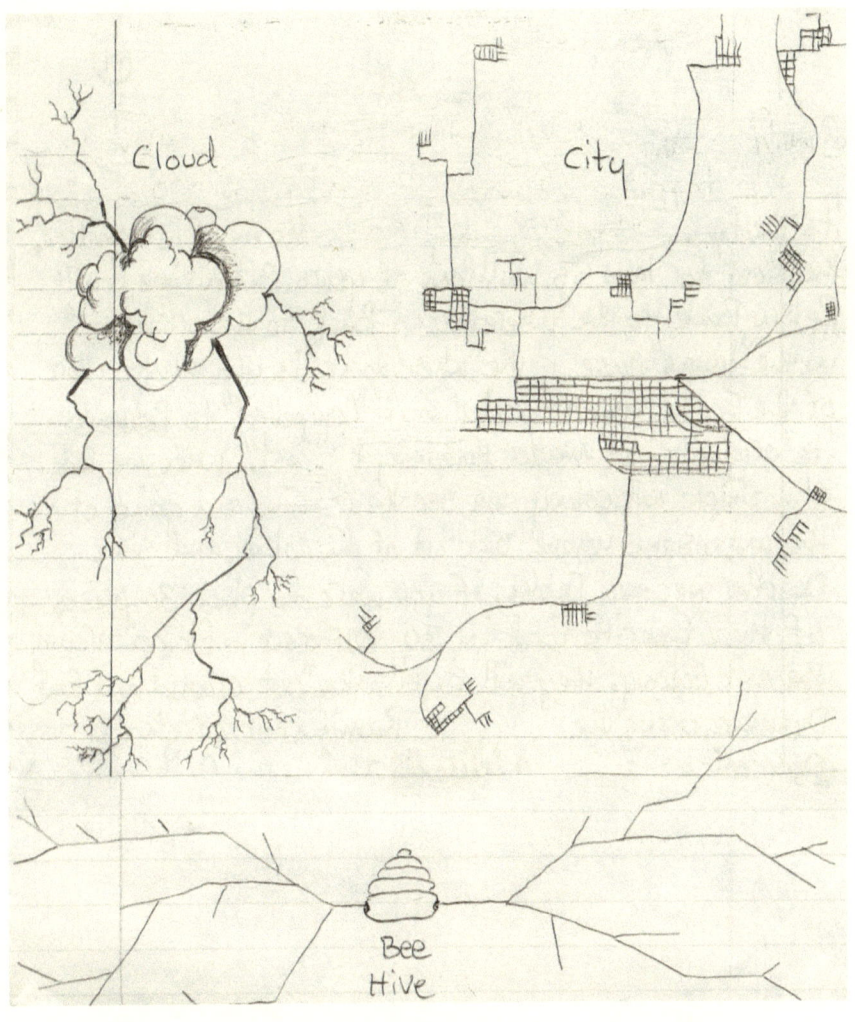

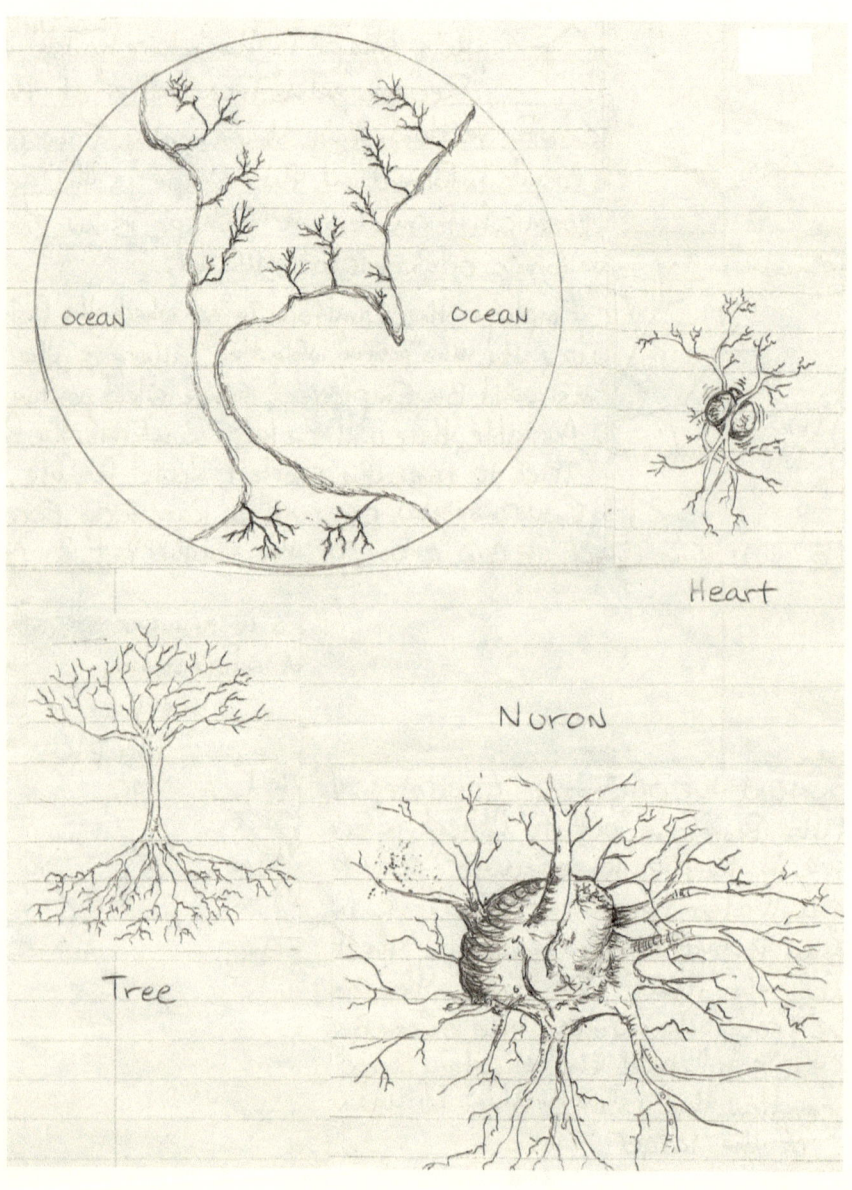

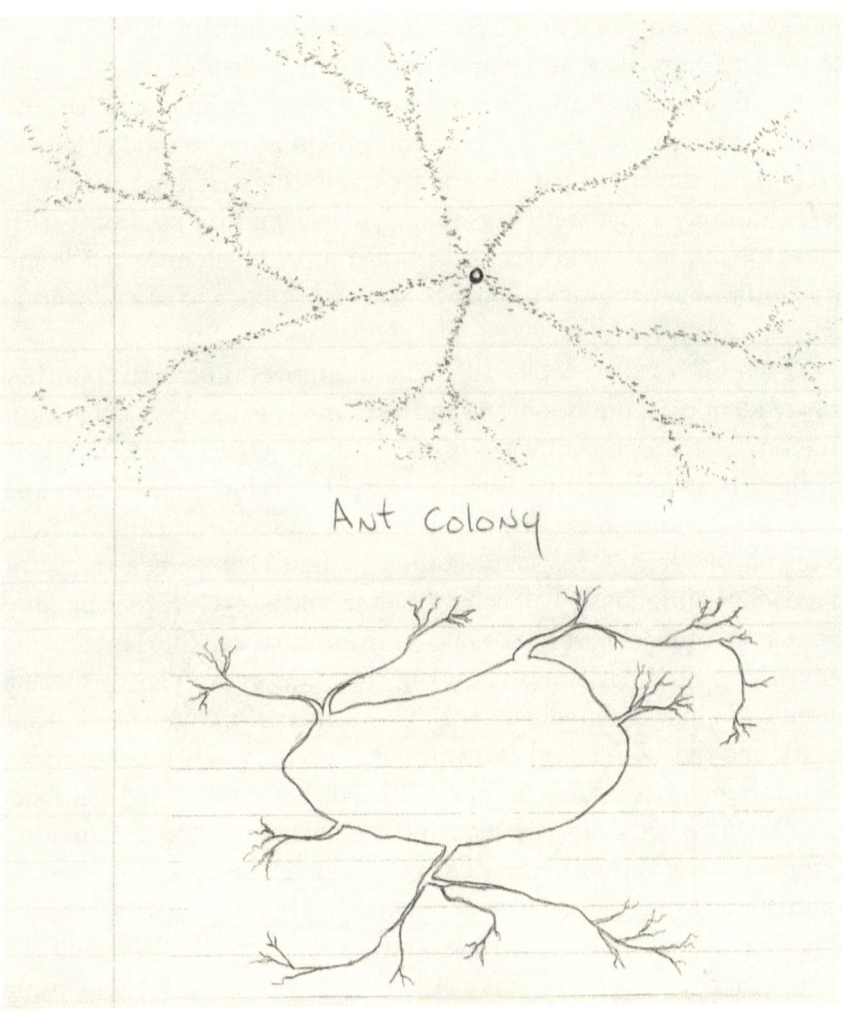

Ant Colony

Is this your liver, or is this a lake? You're right, the answer is "both". Because cosmically, they are the same. Your liver and it's ramiforms provide the same function for matter as a lake and it's ramiforms, and the same as every ant colony and it's ramiforms, and the same as every beehive and it's ramiforms, and every cloud and it's ramiforms, and every advanced human city and it's ramiforms, and every tree and it's ramiforms, and every heart and it's ramiforms, and every ocean and it's ramiforms, and every neuron and it's ramiforms, and every volcano and it's ramiforms, and every glacier and it's ramiforms, and every galaxy and it's ramiforms that move too slow to see, and every solar system and it's ramiforms that move too slow to see, and every molecule, and organic molecule and it's ramiforms that move too fast to see, and every atom and it's ramiforms that move too fast to see, and every atomic particle and it's ramiforms that move too fast to see, and every subatomic particle and it's ramiforms that move too fast to see, every super subatomic particle and it's ramiforms that move too fast to see, and every super super subatomic particle and it's ramiforms that move too fast to see, and every super super super subatomic particle and it's ramiforms that move too fast to see, and so on, and so one, and so on.....................

Each one of these looks different, and moves differently, but they all are identical. Your heart, and the lake, and the tree reveals that each one of these is one complete cosmic "Whole" component. Each one of these is an information warehouse, with specific information, and a specific information transmitting system, and a specific information receiving system. Each specific cosmic "Whole" component transmits a specific, continuously balancing amount, and a specific, continuously balancing sequence of specific nourishing information "into" the specific nourishing information warehouse of the larger encapsulating specific cosmic "Whole" component, that it is a micro component of. Your heart, and the atom reveal that without a specific smaller encapsulated transmission of a specific, continuously balancing amount, and a specific, continuously balancing sequence of specific nourishing information "into" a larger encapsulating cosmic "Whole" component's specific nourishing information warehouse, the larger encapsulating cosmic whole component will deactivate, and disintegrate. The lake, and the neuron, and the tree reveal that each specific smaller encapsulated cosmic "Whole" component receives a specific, continuously balancing amount, and a specific, continuously balancing sequence of specific nourishing information "into" it's specific nourishing information warehouse "from" it's specific larger encapsulating cosmic "Whole" component's specific nourishing information warehouse. The lake, and the neuron, and the tree reveal that without a specific larger encapsulating transmission of a specific, continuously balancing amount, and a specific, continuously balancing sequence of specific nourishing information "into" it's specific smaller encapsulated nourishing information warehouse, the smaller encapsulated cosmic "Whole" component will deactivate, and disintegrate. The volcanoes, and glaciers, and neurons, and trees, and ant colonies, and oceans, and lakes, and beehives, and clouds reveal that a specific, continuously balancing amount, and a specific, continuously balancing sequence of specific replicants behave as one. The neurons, and the ant colonies, and the volcanoes, and the glaciers reveal that the continuous transmission of a specific, continuously balancing amount, and a specific, continuously balancing sequence of specific nourishing information can continuously fluctuate "safely", within a specific window of balance.

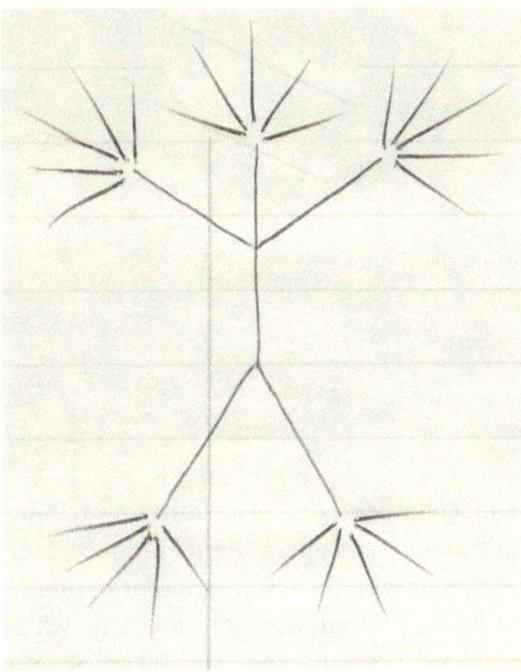

This "shape" is a specific transformation of a ramiformic information receiver, and warehouse, and transmitter. Do you recognize this shape? This is the cosmic "receiver" of matter's information. This is the cosmic "warehouse" of matter's information. This is the cosmic "transmitter" of matter's information.

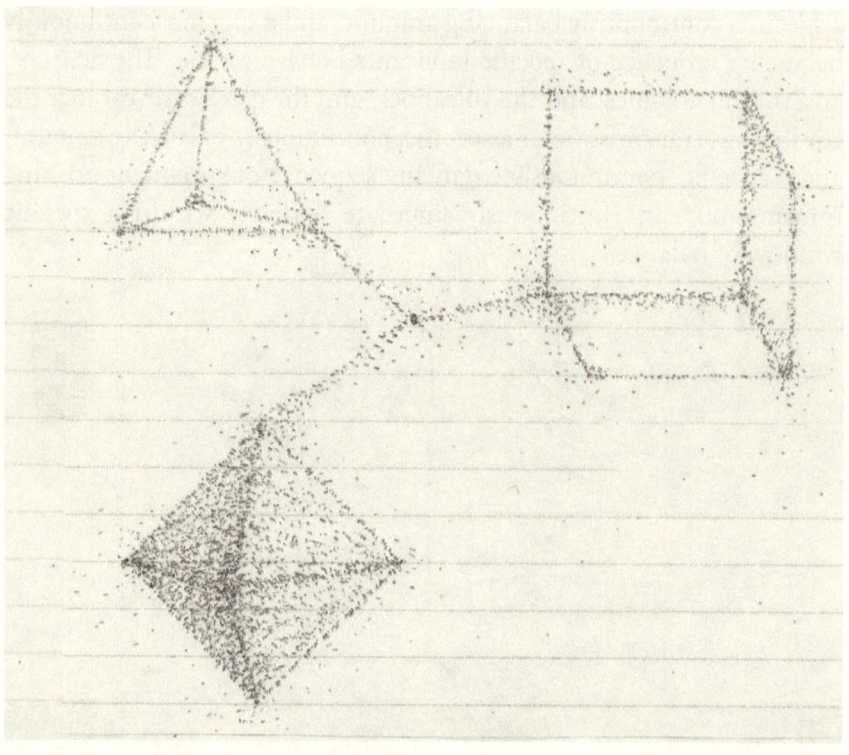

"Only" the ant colony reveals that a ramiform can produce any shape, and maintain any shape indefinitely. Matter produces every straight line, and every specific angle with the ramiform. Matter produces each one of the five prime solids of accumulated atomic matter with the ramiform. Matter produces these five prime specific solid shapes of information to continuously nourish a larger encapsulating cosmic whole component. These five prime specific solid shapes of information, are the specific solid shapes of the four galactic star seeds, and the galaxy activating superstar seed, that when fully processed, will be the components of the post atomic universe's, post atomic neutron. Matter does everything for a reason. Nothing matter does is an accident.

Cosmic Accumulation

Every cosmic "whole" component is an "accumulation" of many smaller cosmic "whole" components. Nothing in our universe "grows". Atomic and molecular accumulations expand and contract, but they do not "grow". A silver atom does not "grow" to the size of a quarter. A balancing accumulation of silver atoms is a quarter.

Every cosmic "whole" accumulation is produced by two cosmic "whole" components "engaging" to produce a rhythmic seed of accumulation regulation activation. In every case of cosmic accumulation, It is the rhythmic seed of accumulation activation that "regulates" the cosmic "whole" component's accumulation result.

Every universe starts from a rhythmic "seed" of activation, then accumulates from the "outside out", encapsulating the "seed" of activation.

Every galaxy activating super star starts from a rhythmic "seed" of activation, then accumulates from the "outside out", encapsulating the "seed" of activation.

Every galaxy starts from a rhythmic "seed" of activation, then accumulates from the "outside out", encapsulating the "seed" of activation.

Every one of the four different post atomic neutron producing galaxy stars, starts from a rhythmic "seed" of activation, then accumulates from the "outside out", encapsulating the "seed" of activation.

Every solar system starts from a rhythmic "seed" of activation, then accumulates from the "outside out", encapsulating the "seed" of activation.

Every Earth, and every other planet, and every planet's moons, starts from a rhythmic "seed" of activation, then accumulates from the "outside out", encapsulating the "seed" of activation.

Every plant, and every insect, and every animal starts from a rhythmic "seed" of activation, then accumulates from the "outside out", encapsulating the "seed" of activation.

Every molecule, and every organic molecule, and every molecular compilation, and every organic molecular compilation, and every cell, and every cellular compilation starts from a rhythmic "seed" of activation, then accumulates from the "outside out", encapsulating the "seed" of activation.

Every element starts from a rhythmic "seed" of activation, then accumulates from the "outside out", encapsulating the "seed" of activation.

Every atom starts from a rhythmic "seed" of activation, then accumulates from the "outside out", encapsulating the "seed" of activation.

Every atomic neutron, and every proton, and every electron, and every photon starts from a rhythmic "seed" of activation, then accumulates from the "outside out", encapsulating the seed of activation.

Every subatomic particle starts from a rhythmic "seed" of activation, then accumulates from the "outside out", encapsulating the "seed" of activation.

Every super subatomic particle starts from a rhythmic "seed" of activation, then accumulates from the "outside out", encapsulating the "seed" of activation.

Every super super subatomic particle starts from a rhythmic "seed" of activation, then accumulates from the "outside out", encapsulating the "seed" of activation, and so on, and so on, and so on....................

Every "accumulating" atomic cosmic whole component is produced with continuously balancing information circulation. Each specific accumulating atomic accumulation continuously circulates a specific, continuously balancing amount, and a specific, continuously balancing sequence of specific smaller matter into, and out of it's overall "system". Each specific accumulating atomic accumulation continuously retains a specific, continuously balancing amount, and a specific, continuously balancing sequence of this continuously circulating, specific, continuously balancing amount, and specific, continuously balancing sequence of specific smaller matter. When each specific atomic cosmic whole component's "accumulating" cycle is over, it continuously stays activated by continuously circulating a specific, continuously balancing amount, and a specific, continuously balancing sequence of specific smaller matter into, and out of it's overall system, and continuously retaining a specific, continuously balancing amount, and a specific, continuously balancing sequence of the continuously circulating specific, continuously balancing amount, and specific, continuously balancing sequence of specific smaller matter, until it deactivates, and disintegrates. This is the "**continuously balancing**" behavior of every universe, every galaxy activating super star, every galaxy, every solar system, every star, every planet, every moon, every animal, every plant, every insect, every cellular compilation, every cell, every organic molecular compilation, every molecular compilation, every organic molecule, every molecule, every element, every atom, every proton, every neutron, every electron, and every photon, and every subatomic particle, and every super subatomic particle, and every super super subatomic particle, and so on, and so on, and so on....................

The Balancing "Systems" Of Matter

Our galaxy is trillions of spacific, continuously balancing systems, and it is a micro component of the universe.

Our earth is trillions of specific, continuously balancing systems, and it is a micro component of the galaxy.

A human is trillions of specific, continuously balancing systems, and it is a micro component of the earth.

A white blood cell is trillions of specific, continuously balancing systems, and it is a micro component of a human.

An organic molecule is trillions of specific, continuously balancing systems, and it is a micro component of a white blood cell.

An atom is trillions of specific, continuously balancing systems, and it is a micro component of an organic molecule.

An atomic particle is trillions of specific, continuously balancing systems, and it is a micro component of an atom.

A subatomic particle is trillions of specific, continuously balancing systems, and it is a micro component of an atomic particle.

A super subatomic particle is trillions of specific, continuously balancing systems, and it is a micro component of a subatomic particle.

A super super subatomic particle is trillions of specific, continuously balancing systems, and it is a micro component of a super subatomic particle, and so on, and so on, and so on.....................

All of these trillions and trillions of specific, continuously balancing systems are simply specific accumulation transformations of the same specific, continuously balancing system. In our universe, there is a specific amount of transformations of this one specific, continuously balancing system. As atomic matter accumulates into larger accumulations, it slows down, and transforms revealing many of these specific transformations. Each specific, cosmic whole component has a specific, continuously balancing amount, and a spacific, continuously balancing sequence of specific accumulation transformations of this one specific, continuously balancing system, producing the "complete" anatomy of "Every" cosmic whole component.

An electron has the identical "complete" anatomy as a mercury atom, and an organic molecule, and a white blood cell, and a tree, and a star, and "**You**". Each one is simply a specifically different, "overall" transformed accumulation motion result. Many atomic cosmic whole components move too slow, or too fast to see, and much of their "complete" anatomy is invisible, even though it is "always" present. Every continuously balancing system that continuously balances "**You**", is also, in a transformed way, exhibited by "every" subatomic particle, and "every" photon, and "every" electron, and "every" proton, and "every" neutron, and "every" atom, and "every" element, and "every" molecule, and "every" organic molecule, and "every" molecular, and organic molecular compilation, and "every" cell, and every cellular compilation, and "every" insect, and "every" plant, and "every" animal, and "every" moon, and "every" planet, and "every" star, and "every" solar system, and "every" galaxy, and "every" galaxy activating super star, and "exactly" the same continuously balancing systems as "every" Universe. Each one is simply a larger, and slower transformed accumulation motion result, or a smaller, and faster transformed accumulation motion result.

The Anatomy of Matter

The complete "anatomy" of every specifically different atomic cosmic whole component is the same. Every atomic cosmic "whole" component's complete "anatomy" is a specific, continuously balancing amount, and a specific, continuously balancing sequence of ramiformicly interconnected, spacific continuously transmitting and recieving, information filters and processors, or translators, encapsulating a ramiforic, spacific continuously transmitting and recieving, information warehouse. In our universe, the atomic cosmic "whole" components are: photons, electrons, protons, neutrons, atoms, elements, molecules, organic molecules, molecular compilations, organic molecular compilations, cells, cellular compilations, plants, animals, insects, planets, moons, stars, solar systems, star clusters, galaxies, galaxy clusters, fully processed galaxy centers, fully processed galaxy center clusters, fully processed galaxy center cluster layers, and our universe itself. Each and every one of these atomic cosmic whole componants has the same complete "anatomy" of continuously transmitting and recieving, information filters and processors that "you" have. Every "behavior" your body exhibits, is exhibited by an electron, and a star, and every single other atomic cosmic "whole" component. They do not watch TV, or laugh with friends, but when you are asleep, "everything" your body is doing, is what a tree's body is doing, and what an insect's body is doing, and what a galaxy's body is doing, and what an earth's body is doing, and it's "exactly" what "every" atom's body is doing.

An atom's body, or complete "anatomy" is very small, and very fast. When atoms accumulate, this accumulation becomes larger, and begins to slow down to become a transformed slow-motion, or suspended animation "graph" of an atom's smaller, and faster complete "anatomy".

At the "**Center**" of every atomic cosmic whole component is "one" central, continuously balancing, rhythmic pulsing continuously transmitting and recieving warehouse of spacific information. Encapsulating this "**Center**", is a "**Middle**".

In the "**Middle**" of every atomic cosmic whole component is a specific, continuously balancing amount, and a specific, continuously balancing sequence of specific temperature regulated, and pressure regulated spacific continuously transmitting and recieving, information filters and processors, or translators. Encapsulating this "**Middle**", is an "**Outside**".

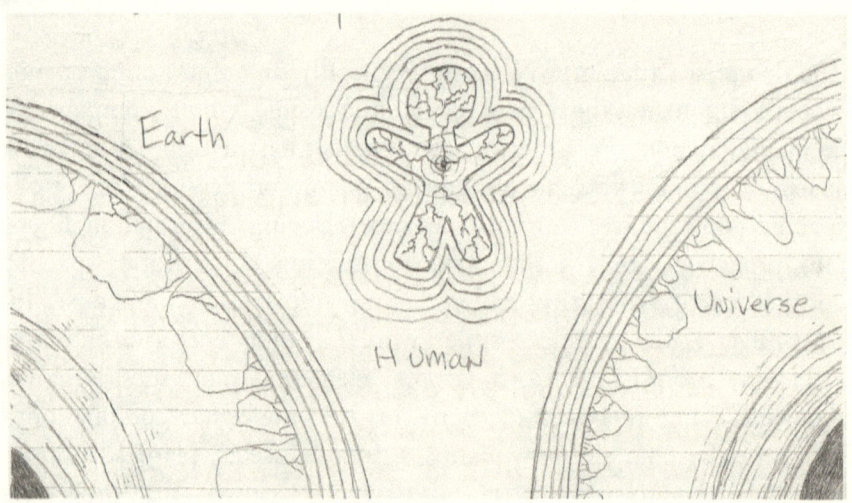

On the "**Outside**" of every atomic cosmic whole component is a specific, continuously balancing amount, and a specific, continuously balancing sequence of specific temperature regulating, and pressure regulating spacific continuously transmitting and recieving, information filters and processors, or translators. The "**Outside**" encapsulates the "**Middle**", and the "**Center**".

The complete anatomy of every atomic cosmic whole component is interconnected by specific transformations of ramiformic activity.

The "outside" of an atomic cosmic "whole" component is what holds it together. The "outside" of an atomic cosmic "whole" component is what attracts it, and connects it to another atomic cosmic "whole" component. The "outside" is also what repels it from other atomic cosmic "whole" components. It also can be neutral to other atomic cosmic whole components, like the Earth's "outside".

78

The "outside" of every atomic cosmic "whole" component is a specific, continuously balancing amount, and a specific, continuously balancing sequence of specific encapsulating "waves". The earth, and the human reveal the most balanced amount, and the most balanced sequence of external encapsulating "waves". The earth, and the human reveal that the external encapsulating "waves" continuously regulate the temperature, and pressure inside of the external encapsulating "waves". They also continuously regulate the size, and shape, and velocity of the matter that continuously enters, and continuously exits an atomic cosmic "whole" component's internal systems.

Every atomic cosmic whole component exhibits a specific, continuously balancing amount, and a specific, continuously balancing sequence of specific transformations of a stratosphere, a troposphere, a mesophere, an ionosphere, and a thermosphere. Like the five outermost sequential epidermic layers of your skin. Every atomic cosmic "whole" component's "skin" exhibits a specific transformation of "pores", like the Earth's volcanoes, or a star's solar flares. Every atomic cosmic "whole" component's "skin" exhibits a specific transformation of continuous "exfoliation". Like a butterfly shedding it's caterpillar, or a crab crawling out of it's old shell, or a snake sneaking out of his old skin, or a uranium atom "continuously" radiating smaller, and faster matter. Like the plastic taste of water, in a plastic bottle. Every atomic cosmic "whole" component's "skin" exhibits a specific transformation of an unexpected matter interaction early warning system, or "feelers". Like the magnetic field, or gravity field of the earth, or the sun. Like the little hairs on micro insects. Like the electric sensors on a shark's nose. Like the hairs on your arm, or leg, or head. Like the hairs on a tomato plant.

Every atomic cosmic whole component exhibits a specific, continuously balancing amount, and a specific, continuously balancing sequence of specific transformations of a nervous system, a digestive system, a skeletal system, a respiratory system, a muscular system, a lymphatic system, an immune system, a circulatory system, and a reproductive system. Just like "one" of your white blood cells, or the earth, or the sun, or a tree, or an insect.

Your nervous system reveals that every specific "system" of an atomic cosmic "whole" component, also has a specific amount, and a specific sequence of specific transformations of the complete "anatomy" of an atomic cosmic whole component. Your "nervous" system has a "skin". It has a digestive system, a skeletal system, an immune system, a muscular system, a circulatory system, a respiratory system, a lymphatic system, and a reproductive system. All of these systems reveal that each, and every one of them is a full complement of each and every other one, even itself. Every immune system, has an immune system. Every skeletal system, has a skeletal system, every nervous system, has a nervous system, every muscular system, has a muscular system, and every specific reproductive system, has a specific reproductive system.

Our sun reveals that a star's respiratory system, is also it's circulatory system. A star's endocrine system, is also it's muscular system. A star's skin, is also it's reproductive system. A star's nervous system, is also it's skelital system, and a star's digestive system, is also it's immune system. Our sun's continuous behavior is each one, and every one of these specific systems.

Our sun also reveals that each, and every specific system of atomic matter, is simply a specific accumulation transformation of "**one**" specific system. This specific system is the system of "**Continuously Balancing Information Circulation**", and of course, every specific transformation of "continuously balancing information circulation" is simply a specific, continuously balancing accumulation transformation of the journey of matter. "Out half around rest, half around back rest".

Digestion is how atomic matter cultures it's maximum complexity retentive, "maximum direct energy output %, to indirect energy output %, in relation to overall energy input", continuously balancing atomic components. Matter is simply in a perpetual state of processing it's smaller self, into it's larger self.

Matter is in a continuous state of "circulation".
Matter is in a continuous state of "calculation".
Matter is in a continuous state of "accumulation".
Matter is in a continuous state of "disintegration".

Every atomic cosmic whole component "ingests", or "eats" a specific, continuously balancing amount, and a specific, continuously balancing sequence of specific "smaller" cosmic "whole" components. Just like the galcleous eats a specific, continuously balancing amount, and a specific, continuously balancing sequence of smaller dead stars, and radiant stars, and planets, and moons. Just like the sun eats a specific, continuously balancing amount, and a specific, continuously balancing sequence of smaller hydrogen atoms, and space dust, and comets and asteroids. Just like the Earth eats a specific, continuously balancing amount, and a specific, continuously balancing sequence of smaller incoming space elements, and atoms, and molecules, and asteroids, and comets. Just like you ingest a specific, continuously balancing amount, and a specific, continuously balancing sequence of smaller atoms, and molecules when you continuously breathe, and you periodically drink liquid, and eat food. Just like an ant breathes in a specific, continuously balancing amount, and a specific, continuously balancing sequence of smaller air, and periodically drinks and eats a specific, continuously balancing amount, and a specific, continuously balancing sequence of smaller food. Just like a white blood cell ingests a specific, continuously balancing amount, and a specific, continuously balancing sequence of smaller molecules. Just like a uranium atom ingests, or "eats" a specific, continuously balancing amount, and a specific, continuously balancing sequence of smaller cosmic "whole" unified subatomic particle clusters.

Every atomic cosmic whole componants continuously transmitting and receiving information warehouse, continuously calculates the continuously balancing influx of smaller specific cosmic whole components.

Every atomic cosmic whole component "Emits" a specific, continuously balancing amount, and a specific, continuously balancing sequence of direct energy, and a specific, continuously balancing amount, and a specific, continuously balancing sequence of indirect energy, according to it's specific information warehouse's continuous, specific calculation of the continuously balancing, incoming specific smaller cosmic whole components. Just like the galcleous's continuously balancing direct energy output % is it's continuous physical behavior, and it's continuously balancing indirect energy output % is it's "Jets" of vaporized stars, and planets. Just like the Sun's continuously balancing direct energy output % is it's continuous physical behavior, and it's continuously balancing indirect energy output % is approximately four hundred million tons of helium per second. Just like the Earth's continuously balancing direct energy output % is it's continuous physical behavior, and it's continuously balancing indirect energy output % is tons and tons of "gases" per second. Just like your continuously balancing direct energy output % is your continuous physical behavior, and your continuously balancing indirect energy output % is your continuous exhalation, transpiration, exfoliation, and periodic excretion. Just like your white blood cell's continuously balancing direct energy output % is it's continuous physical behavior, and it's continuously balancing indirect energy output % is thousands, or millions of organic molecules per second. Just like a uranium atom's continuously balancing direct energy output % is it's continuous physical behavior, and it's continuously balancing indirect energy output % is millions, or billions of unified subatomic particle clusters per second.

During each atomic cosmic whole component's specific cosmic lifespan, it incorporates, or retains, or accumulates a specific, continuously balancing amount, and a specific, continuously balancing sequence of the continuously balancing incoming, and outgoing smaller specific cosmic whole components, in accordance with the continuous calculation of it's specific information warehouse.

At the end of every atomic cosmic whole component's complete cosmic lifespan, it will deactivate, and disintegrate.

At a specific point during each atomic cosmic whole component's complete, continuous, specific smaller cosmic whole component, specific, continuously balancing amount, and specific, continuously balancing sequence circulation cycle, or cosmic lifespan, it will leave behind a very small specific accumulation result of it's specific information warehouse's continuous calculations during it's specific, complete, continuous, specific smaller cosmic whole component, specific, continuously balancing amount, and specific, continuously balancing sequence circulation cycle, or cosmic lifespan. Just like the center of our galaxy, will be the smaller accumulation result of our entire galaxy. Just like a white dwarf star's super compressed, spherical diamond, star seed core, will be the smaller accumulation result of our sun. Just like the smaller human-like creature, and the diamond, is the accumulation result of our earth. Just like the staminate, and the pistilate is the smaller accumulation result of the oak tree. Just like the smaller egg, or sperm, is the accumulation result of you. Just like DNA, is the accumulation result of your white blood cell, and just like many smaller atoms, or many smaller atomic particles, is the accumulation result of every uranium atom.

The anatomy of a tree reveals the transformed, suspended animation graph of the "complete" anatomy of every specific atomic cosmic whole component. When we look at a cross section of a one hundred year old tree, we see one hundred encapsulating accumulation transformations, of one hundred cycles, of primarily the exact same information.

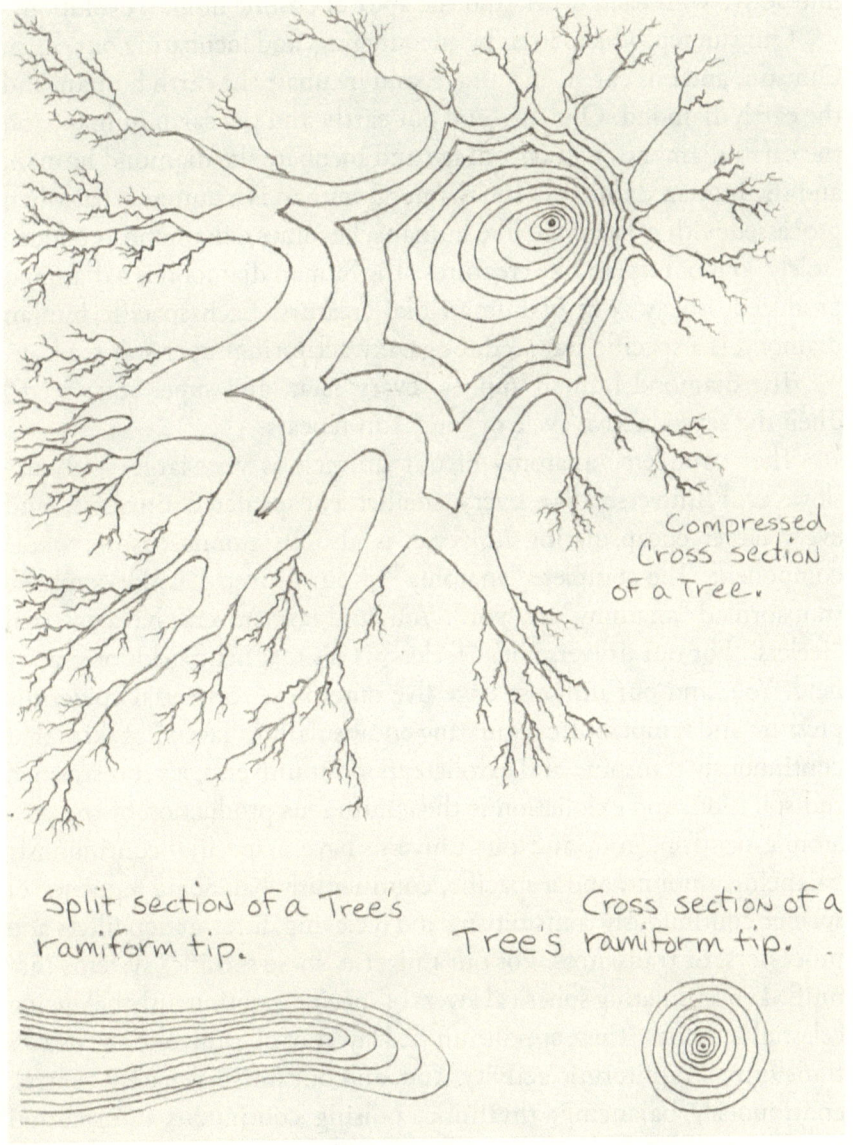

Compressed Cross section of a Tree.

Split section of a Tree's ramiform tip.

Cross section of a Tree's ramiform tip.

Each encapsulating advancing layer is a larger, and slightly transformed accumulation replication of each smaller encapsulated layer. The "Tree" reveals that every atomic cosmic whole component's anatomy is endless smaller encapsulated transformed layers of specifically different amounts less of the exact same information.

Every atomic cosmic whole component has a specific reproduction system, and they are "all" continuously interconnected and continuously interactive with each other, and the apex of cosmic human evolution.

Our sun reproduces itself by encouraging, and incubating our earth. Our sun, and our earth, encourage, and incubate the earth human, and the earth diamond. Our sun, and our earth, and the earth human, and the earth diamond, will encourage and incubate the diamond human, and the human diamond. The diamond human is a human, that when processed with other same like humans, becomes a diamond produced "solely" out of human like creatures. The human diamond is a diamond produced "solely" out of human like creaturs. Each spacific human diamond is a specific star seed, or galaxy activating super star seed.

The diamond human applies "every" star, and super star "seed". Then the reproductive cycle of "our" sun repeats.

The "complete" anatomy of our universe is very large, and very slow. Our universe, like every smaller encapsulated universe, and every larger encapsulating universe, is also an atomic cosmic whole component. The complete "anatomy" of our universe, is the complete transformed "anatomy" of "**you**". You, and our universe have external "feelers". For our universe, it's "feelers" is it's magnetic field, or gravity field. You, and our universe have five outermost sequential epidermic pressure and temperature regulating encapsulating "layers" of skin, that continuously transpire, and exfoliate. For our universe, it's transformed transpiration, and exfoliation is the continuous production of the post atomic neutron. You, and our universe have a specific, continuously balancing amount, and a spacific, continuously balancing sequence of specific continuously transmitting and recieving, information filters and processors, or translators. For our universe, these specific "systems" are unified encapsulating spherical layers of specific, continuously balancing celestial activities. These specific unified layers of our universe are also it's transformed ramiformic activity. You, and our universe have a central, continuously balancing, rhythmic, pulsing continuous information transmitter, receiver, and warehouse of continuously balancing outgoing specific information, and continuously balancing incoming specific information. You, and our universe "continuously" pump three smaller specific information warehouses throughout it's complete anatomy. Each specific warehouse will leave a specific, continuously balancing amount, and a specific, continuously balancing sequence of it's specific information, in a specific, continuously balancing amount, and a specific, continuously balancing sequence of specific areas. Then each of the three will return back to the information "pump". Each of the three are now smaller from dropping off their specific information. After returning back to the information pump, or generator, each of the three smaller information warehouses are restocked with the same information, and sent right back out into the complete anatomy. For "you", the central, continuously balancing, rhythmic, pulsing continuous information transmitter, receiver, and warehouse is your heart. For our universe, it is the ucleous, or subatomic universe. For you, the three specific smaller, continuously, rhythmically, exiting warehouses of information are the red blood cells, the white blood cells, and the leukocytes. For our universe, the proton, the electron, and the neutron will continuously deliver their information indefinitely, unless they are "**shaken**" apart into subatomic neutrons. These subatomic neutrons "continuously", "rhythmically", return back to the ucleous. The subatomic universe.

The human, and our universe reveal that "**Every**" atomic cosmic whole component, is a specific transformation of each and "**Every other**" atomic cosmic whole component.

The Sphere

The sphere "pattern", or the sphere "graph" represents the most balanced accumulations of atomic matter, and matter itself. Every cosmic sphere's cosmic life span is based on it's continuously balancing "size". The more cosmically balanced it's "size" is, the longer it's cosmic lifespan is. The "size" of a star that is one and a half times bigger than our Sun is very balanced. This "size" sphere can last up to ten million years. The specific sphere size of the star that is two times, and three times the specific sphere size of the sun are also very balanced, and can also exhibit a cosmic life span of up to ten million years. The specific sphere size of our sun, and our earth, and our moon, and "all" of our solar system's planets, and moons exhibit a specific sphere size of balance. Our sun, and our earth, and our moon exhibit three of our universe's most balanced specific sphere sizes. Each one has a cosmic life span that can last up to ten billion years.

The specific sphere size of a human's white blood cell is also one of our universe's most balanced sphere sizes. This specific, continuously balancing size can last indefinitely. Just like the specific, continuously balancing spherical size of a proton, or neutron, or an electron can last indefinitely. When a proton, or neutron's indefinitely, continuously balancing spherical size "changes" inside of an atom, or molecule, or element, or molecular compilation, or organic molecular compilation, it is simply expanding, or contracting into one of a very large specific amount of our universe's most balanced spherical sizes. The complete specific amount, and complete specific sequence of our universe's most balanced spherical sizes, is the transformed information of every "frequency", or complete circle cycle of the electromagnetic spectrum.

The Substance Of Matter

Our universe is a cosmic "substance" assembly factory. Our universe does not "create" the cosmic substance, it simply reassembles it around it's smaller and faster identical substance self. Our universe is using the same substance, and the same formula to produce the post atomic neutron, as the subatomic universe uses to produce the atomic neutron. As we follow the accumulation results of the atomic neutron, matter reveals what it's smaller and faster particles of "substance" are.

Atomic matter only does "one" thing. It continuously reproduces itself. Matter is only made up of "one" substance. Itself.

The journey of the atomic neutron starts at the subatomic universe. The atomic neutron is the maximum complexity retentive finished product of the subatomic universe. Matter itself is a maximum complexity retention system. Cosmic complexity is based on three things. Cosmic life span, cosmic mobility, and cosmic knowledge. The atomic neutron is the longest living, most complex moving, most intelligent component in our universe, and our universe is continuously producing a larger and slower replicant of the atomic neutron.

When we look at our galaxy, we are looking at a cosmic substance processor. Cosmically, our galaxy is very young. Our galaxy is merely an embryo of it's future self. Our Milky Way galaxy is an accumulation of "many" previously separate galaxies existing as one. The Andromeda galaxy is heading directly toward our galaxy at three hundred thousand miles per hour. When it arrives, these two galaxies will it exist as one. The center of our galaxy is the nucleus of our galaxy. The nucleus of our galaxy is commonly known as a "Black Hole". Our galaxy's nucleus is not "black", it is a specific, continuously balancing amount, and specific, continuously balancing sequence of very small "**Clear**" spheres. Our galaxy's nucleus is not a "Hole". It is a specific, continuously balancing amount, and a specific, continuously balancing sequence of components of cosmic intention. There are no holes in space. There is no vortex to another dimension, just a nucleus of cosmic intention.

Our galaxy's nucleus is "continuously" accumulating. For our galaxy's nucleus to accumulate properly, it must be fed a specific, continuously balancing diet. Our galaxy's nucleus ingests a specific, continuously balancing amount, and a specific, continuously balancing sequence of specific dead stars, and radiant stars, and planets, and moons. Then it retains the intended components, and shoots the unintended, or waste components out in jets of vaporized particles. The cosmic lifespan of our galaxy's nucleus is continuously increasing. The cosmic mobility complexity of our galaxy's nucleus is continuously increasing. The warehouse of cosmic knowledge, that is our galaxy's nucleus, is continuously increasing. It's cosmic life span, and cosmic mobility, and cosmic knowledge will continue to increase until the nucleus of our galaxy becomes part of the longest living, most complex moving, most intelligent cosmic component that our universe produces.

When our galaxy's nucleus is finished accumulating, it will be a specific, continuously balancing amount, and a specific, continuously balancing sequence of five specifically different sized, fully processed, star "seeds". Each star seed begins as a "human diamond", and it is heated and compressed until it is fully processed. Then it is collected at the galaxy nucleus. These five specific, fully processed star seeds are the "substance" of matter.

The "substance" of matter is a specific, continuously balancing amount, and a specific, continuously balancing sequence of "five" specifically different sized, rotating, wobbling, flipping, quaking, and orbiting, spherical, super compressed, super dense, super small, super fast, super attractive, super clear, super "**Diamonds**".

Our galaxy evolves into a galaxy of super diamonds. This specific super diamalaxy, and a specific, continuously balancing amount, and a specific, continuously balancing sequence of other specific super diamalaxys, accumulate to produce the cosmic super diamalaxsphere. The cosmic super diamalaxsphere, is the cosmic neutron.

The substance of matter, is

"Super Diamonds".

The Sound Of Matter

Every photon, and electron, and proton, and neutron is filled with continuously balancing rotating, and wobbling, and flipping, and quaking, and orbiting particles. The continuously balancing behavior of these particles are responsible for the continuously balancing motion results that produces each atomic particle's continuously balancing external behavior. Each specific atomic particle's continuously balancing external behavior is simply the result of a specific, continuously balancing amount, and a specific, continuously balancing sequence of specific, continuously balancing repetitive "**sound cycles**".

Our earth, and our moon's continuously balancing repetitive behavior is the exact same behavior of every universal replicant earth, and replicant moon. This specific, continuously balancing behavior is the "key" to atomic matter's accumulating behavior. When the "frequency" of the Earth's specific, continuously balancing rotating, and wobbling, and flipping, and quaking, and orbiting, and the specific, continuously balancing behaviors of the moon are accelerated to the "key" velocity of our universe, they produce the "key" sound of our universe. They produce the "key" tone of our universe. They produce the "key" note of our universe.

Specific, continuously balancing repetitive "sound cycles", is exactly what matter is. Even a super compressed, super diamond is not a solid particle, it is many smaller components producing specific, continuously balancing repetitive "sound cycles". Matter itself is specific, continuously balancing amounts, and specific, continuously balancing sequences, of specific, continuously balancing repetitive sound cycles.

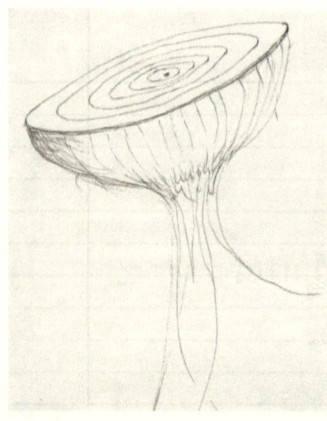

Every cosmic whole component is a specific, continuously balancing amount, and a specific, continuously balancing sequence of specific, continuously balancing repetitive sound cycles. Every specific sound, in every specific, continuously balancing sound amount sequence cycle, is also a specific, continuously balancing amount, and a specific, continuously balancing sequence of specific, continuously balancing repetitive sound cycles, and so on, and so on, and so on....................

The most balanced accumulations of atomic "sounds" are the "spheres".

Every cosmic whole component is a transformed encapsulation, of endless transformed encapsulations. The sphere is simply the most balanced. The sphere is simply a suspended animation graph of the continuous behavior of the most balanced smaller and faster molecular, and atomic compilations. The sphere reveals that every cosmic whole component is specific, continuously balancing amounts, and specific, continuously balancing sequences of specific, continuously balancing repetitive **"sound cycles"**.

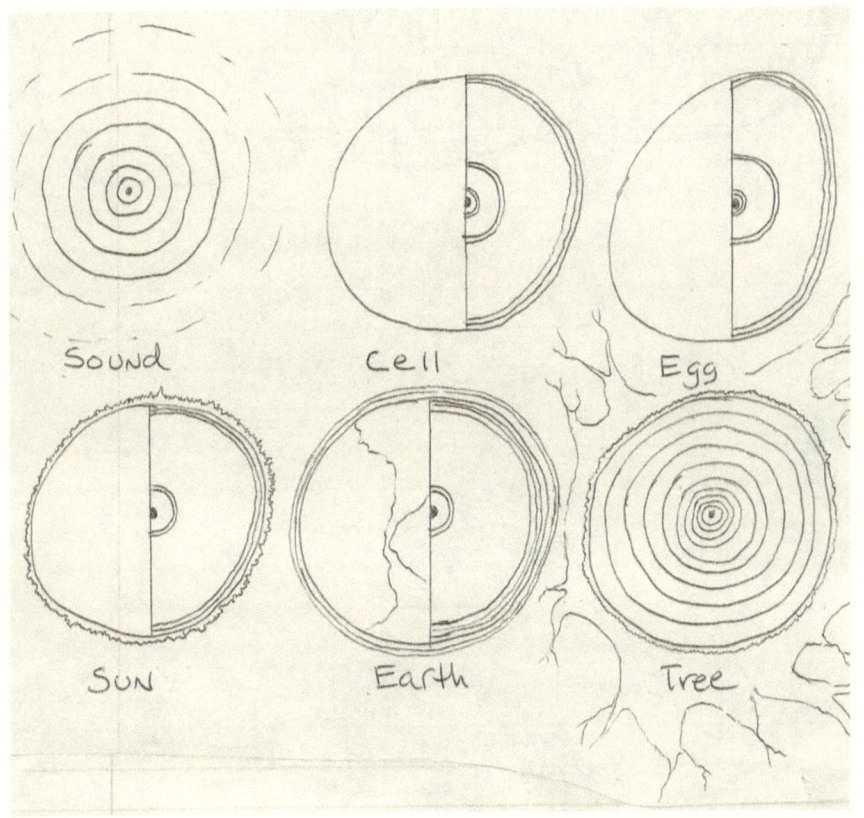

Sound   Cell   Egg

Sun   Earth   Tree

Matter itself, is nothing more, than "**Sound**".

The Balancing Amounts
And Sequences Of Matter

"All" of matter's continuously balancing behaviors are based on specific, continuously balancing repetitive amounts, and specific, continuously balancing repetitive sequence cycles. Every cosmic whole component's continuously balancing behavior is based on a specific, continuously balancing amount, and a specific, continuously balancing sequence of specific "smaller" continuously balancing systems. These specific "smaller" systems are also cosmic whole components, and their continuously balancing behavior is based on a specific, continuously balancing amount, and a specific, continuously balancing sequence of specific "even smaller" continuously balancing systems. These specific "even smaller" systems are also cosmic whole components, and their continuously balancing behavior is based on a specific, continuously balancing amount, and a specific, continuously balancing sequence of specific "even smaller" continuously balancing systems, and so on, and so on, and so on...................

"Every" one, of "Every" atomic cosmic whole component's, specific smaller atomic cosmic whole component's balancing system's, continuously balancing behavior, is based on a continuously oscillating, continuous, repetitive, specific amount, and specific sequence cycle of "stability", centered around a continuous, repetitive, specific amount, and specific sequence cycle of "balance".

Every continuous behavior of our sun, is one of it's specific, continuously balancing, smaller atomic systems. The sun's continuous orbit. The sun's continuous rotation. The sun's continuous quaking.

The sun's continuous wobbling. The sun's continuous flipping. The sun's continuous core temperature. The sun's continuous internal temperature. The sun's continuous external temperature. The sun's continuous core pressure. The sun's continuous internal pressure, and the sun's continuous external pressure are "all" a continuously oscillating, continuous, repetitive, specific amount, and specific sequence cycle of "stability", centered around a continuous, repetitive, specific amount, and specific sequence cycle of "balance".

Some of the sun's continuously balancing, smaller atomic systems are very precise. Just like sun's continuous rotation, and the sun's continuous orbit around the galcleous. Some of the sun's continuously balancing, smaller atomic systems exhibit an obvious continuously oscillating window of stability, centered around a neutral point of balance. Just like the sun's continuous internal temperature, or the sun's continuous flipping behavior. "All" of our sun's specific amount, and specific sequence of specific, continuously balancing smaller atomic systems, are continuously interconnected, and continuously interactive. If "any" of our sun's specific, continuously balancing, smaller atomic systems exceed their individual specific window of stability, then transformation will occur.

Every continuous behavior of our Earth, is one of it's specific, continuously balancing, smaller atomic systems. The Earth's continuous orbit. The Earth's continuous rotation. The Earth's continuous wobbling. The Earth's continuous quaking. The Earth's continuous flipping. The Earth's continuous core temperature. The Earth's continuous internal temperature. The Earth's continuous external temperature. The Earth's continuous core pressure. The Earth's continuous internal pressure. The Earth's continuous external pressure. The Earth's continuous atmospheric behavior. The Earth's continuous oceanic behavior. The Earth's continuous volcanic behavior, the Earth's continuous glacial behavior, and the Earth's continuous biological behavior are "all" a continuously oscillating, continuous, repetitive, specific amount, and specific sequence cycle of "stability", centered around a continuous, repetitive, specific amount, and specific sequence cycle of "balance". Some of the Earth's continuously balancing, smaller atomic systems are very precise. Just like the Earth's continuous orbiting, and rotation around the sun. Some of the Earth's continuously balancing, smaller atomic systems exhibit an obvious specific, continuously oscillating window of stability, centered around a neutral point of balance. Just like the Earth's continuous flipping behavior, and the Earth's continuous volcanic behavior. "All" of our Earth's specific amount, and specific sequence of specific, continuously balancing, smaller atomic systems, are continuously interconnected, and continuously interactive. If "any" of the Earth's specific, continuously balancing, smaller atomic systems exceed there individual spacific window of stability, then transformation will occur.

Every continuous behavior of "You", is one of your specific, continuously balancing, smaller atomic systems. "Your" continuous core temperature. "Your" continuous internal temperature. "Your" continuous external temperature. "Your" continuous core pressure. "Your" continuous internal pressure. "Your" continuous external pressure. "Your" continuous eating behavior. "Your" continuous physical behavior. "Your" continuous sleeping behavior. "Your" continuous respirating behavior. "Your" continuous transpiring behavior. "Your" continuous cell production behavior, and "Your" continuous circulating behaviors are "all" a continuously oscillating, continuous, repetitive, specific amount, and specific sequence cycle of "stability", centered around a continuous, repetitive, specific amount, and specific sequence cycle of "balance". Some of your continuously balancing, smaller atomic systems are very precise, like your continuous cell production, and your continuous blood cleansing. Some of your continuously balancing, smaller atomic systems exhibit an obvious continuously oscillating window of stability, centered around a neutral point of balance. Just like your continuous internal temperature, and your continuous eating behavior. "All" of "your" specific amount, and specific sequence of specific, continuously balancing, smaller atomic systems are continuously interconnected, and continuously interactive. If "any" of "your" specific, continuously balancing, smaller atomic systems exceed there individual spacific window of stability, then transformation will occur. This continuously balancing dynamic, is the identical, transformed behavior of every universe, and every galaxy, and every solar system, and every star, and every planet, and every moon, and every animal, and every plant, and every insect, and every cellular compilation, and every cell, and every molecular compilation, and every organic molecular compilation, and every organic molecule, and every molecule, and every element, and every atom, and every atomic particle, and every subatomic particle, and every super subatomic particle, and every super super subatomic particle, and every super super super subatomic particle, and so on, and so on, and so on....................

The Human cosmic whole component reveals that when one specific, continuously balancing, smaller atomic system accelerates, they "all" accelerate. When one specific, continuously balancing, smaller atomic system decelerates, they all decelerate.

The human cosmic whole component reveals that not only is every specific cosmic whole component's specific amount, and specific sequence of specific, continuously balancing, smaller atomic systems continuously interconnected, and continuously interactive, but they are also continuously interconnected, and continuously interactive with a specific amount, and a specific sequence of specific "larger" cosmic whole components, and their continuously balancing systems, and a specific amount, and a specific sequence of specific "smaller" cosmic whole components, and their continuously balancing systems. The continuously balancing systems of temperature and pressure reveal the continuous interconnection, and the continuous interaction between "all" matter.

The human cosmic whole component reveals that the ucleous's continuously balancing temperature and pressure continuously helps to regulate the continuously balancing temperature and pressure of our galcleous. The continuously balancing temperature and pressure of the galcleous continuously helps to regulate the continuously balancing temperature and pressure of our sun. The continuously balancing temperature and pressure of our sun continuously helps to regulate the continuously balancing temperature and pressure of the earth. The continuously balancing temperature and pressure of the earth continuously helps to regulate "your" continuously balancing temperature and pressure. "Your" continuously balancing temperature and pressure continuously helps to regulate the continuously balancing temperature and pressure of your molecules. The continuously balancing temperature and pressure of your molecules continuously helps to regulate the continuously balancing temperature and pressure of your atoms. The continuously balancing temperature and pressure of your atoms continuously helps to regulate the continuously balancing temperature and pressure of your atomic particles. The continuously balancing temperature and pressure of your atomic particles continuously helps to regulate the continuously balancing temperature and pressure of your subatomic particles, and the continuously balancing temperature and pressure of your subatomic particles continuously encourages the continuously balancing expected response from the Deep Torsion Constant.

The human cosmic whole component reveals that specific larger and slower cosmic whole components continuously encourage a specific response from specific "Middle" cosmic whole components, and specific smaller and faster cosmic whole components are continuously calculating this continuous encouragement, or stimulation, and continuously responding.

The human cosmic whole component reveals that every specific cosmic whole component is a specific translator between the continuous, specific encouragement, or stimulation, of specific larger and slower cosmic whole components, and the continuous, specific calculation and response from specific smaller and faster cosmic whole components.

The human cosmic whole component reveals that every cosmic whole component is a "Middle" cosmic whole component. The human is the result of the continuous encouragement, or stimulation, from larger and slower matter, "**And**" the continuous calculation and response from smaller and faster matter. The human reveals that all matter, is "middle" matter.

When atomic particles become the "Sun", they become a specific, continuously balancing amount, and a specific, continuously balancing sequence of continuously balancing behaviors.

When atomic particles become a "Human", they become a specific, continuously balancing amount, and a specific, continuously balancing sequence of continuously balancing behaviors.

When atomic particles become the "Earth", they become a specific, continuously balancing amount, and a specific, continuously balancing sequence of continuously balancing behaviors.

Our Earth, and every other Earth's continuously balancing behavior, is the "key" to our universe's continuously balancing behavior. The "key" to our earth's continuously balancing behavior is the "key" continuously oscillating, continuous, repetitive, "stabilizing" specific amount, and specific sequence cycle of matter, centered around the "key" continuous, repetitive, "balancing" specific amount, and specific sequence cycle of matter, which is 365.25 and so on, and so on, and so on....................

"All" of the continuously balancing behaviors of accumulated atomic matter are based on continuously oscillating, continuous, repetitive, spacific amounts, and spacific sequence cycles of "stability", centered around continuous, repetitive, spacific amounts, and spacific sequence cycles of "balance. "All" of them have a specific importance. "All" of them are necessary for the accumulation of atomic matter. This is the "base" continuously oscillating, continuous, repetitive, "stabilizing amount sequence cycle of matter", centered around the "base" continuous, repetitive, "balancing amount sequence cycle of matter". 000. 0000000000000 and so on, and so on, and so on.................... These specific amounts, in this specific sequence, are the "Base Amounts" of matter. 365.25

Three

"Three" is the "base" amount of matter. "Three" is the "base" amount, in the "base amounts sequence" of matter. "Three" is the "base" amount, of "base amounts" in the "key" continuous, repetitive, "balancing" specific amount, and specific sequence cycle of matter, that the "key" continuously oscillating, continuous, repetitive, "stabilizing" specific amount, and specific sequence cycle of matter is centered around. "Three" is simply the "prime" amount of matter. Every atomic cosmic whole component has a specific smaller cosmic whole component amount that is a multiple of "three", or can be divided by "three". That's because every photon, and every electron, and every proton, and every neutron's subatomic particle amount is a multiple of "three", or can be divided by "three". That's because every subatomic particle's, super subatomic particle amount is a multiple of "three", or can be divided by three. That's because every super subatomic particle's, super super subatomic particle amount is a multiple of "three", or can be divided by "three", and so on, and so on, and so on....................

Accumulated atomic matter continuously repeats the "graph", or "pattern" of "three".

Accumulated atomic matter continuously repeats the "amount" of "three".

There are "three" base atomic particles.

Proton   Neutron   Electron

There are "three" states of electric.

Positive	Neutral	Negative

There are "three" states of magnetism.

Attractive	Neutral	Repulsive

There are "three" states of biological balance.

Health	Wisdom	Strength

There are "three" states of a biological cycle.

Born	Alive	Dead

There are "three" states of a star's cycle.

Activated	Active	Deactivated

There are "three" states of a cosmic whole component.

Center	Middle	Outside

There are "three" states of cosmic complexity.

Mobility	Intelligence	Lifespan

There are "three" base states of motion.

Out	Around	Back

There are "three" states of dimension.

Height	Length	Width

There are "three" states of heart function.

Intake	Rest	Outflow

There or "three" states of blood information.

White Blood Cell   Leukocytes   Red Blood Cell

There are "three" states of information circulation.

Receiving   Warehousing   Transmitting

There are "three" states of a DNA helix.

Super Coiled   Relaxed   Negative Super Coiled

There are "three" states of music.

Melody   Harmony   Rhythm

The amount of "three" is the "prime" amount, in the "key" continuous, repetitive, "balancing" specific amount, and specific sequence cycle of matter, that the "key" continuously oscillating, continuous, repetitive, "stabilizing" specific amount, and specific sequence cycle of matter is centered around. The amount of "three" is the "prime" amount of "cosmic balance". The amount of "three" is the "prime" amount of matter itself.

Six

The amount of "six" is the "key" amount, in the "key" continuous, repetitive, "balancing" amount sequence cycle of matter, that the "key" continuously oscillating, continuous, repetitive, "stabilizing" specific amount, and specific sequence cycle of matter is centered around. This "key" continuously balancing behavior is what continuously encourages every Earth's continuous behavior. Every Earth's continuous behavior, is simply to continuously produce human like creatures, and to continuously produce diamond, and "carbon" is the "key". Every earth is a human, and a diamond factory, and "carbon" is the "key". "Six" protons. "Six" neutrons. "Six" electrons. The amount of "six" is represented by the "cube". The "cube" is the "key" shape, or solid of matter. The "cube" is the shape of every diamond, star seed, of a star that is three times the mass of the sun. The "frequency", or the "tone", or the "note", or the "sound" that an uncompromised carbon atom continuously makes, is the "key" sound of matter. The "cube" has one angle. This "angle" is the "key" angle of matter.

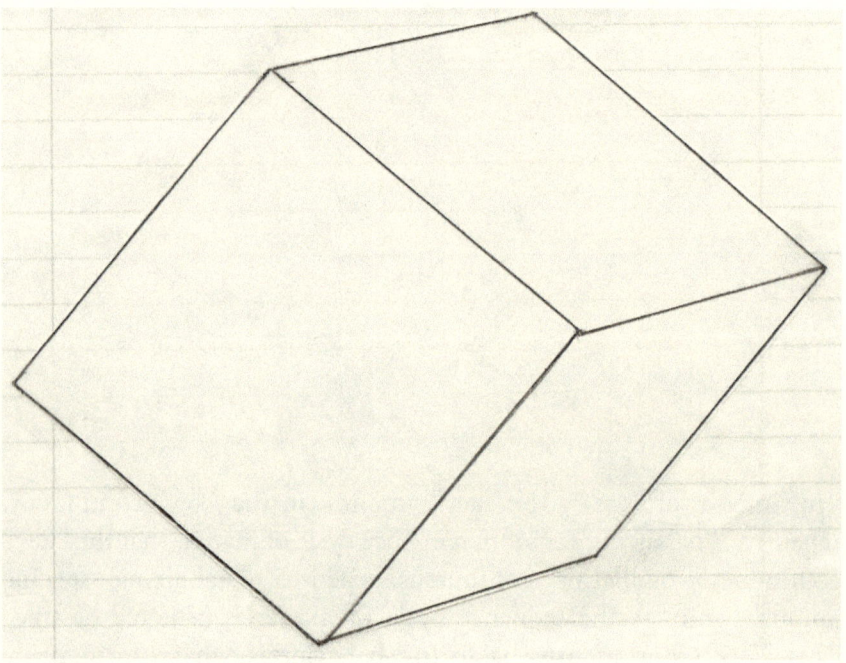

The amount of "six" is the "key" amount of cosmic balance. The amount of "six" is the "key" amount of matter itself.

Five

The amount of "five" is the "constructing" amount of matter. "Five" is the "regulating" amount of matter. "Five" is the "balancing" amount of matter. "Five" is the amount of matter's "structural soundness". Every cosmic whole component is continuously constructed, and regulated, and balanced, and kept structurally sound, or held together by the accumulation result of the amount of "five". Just like the specific sequence of the "five" specific uniospheres of our universe. Just like the spacific sequence of the "five" spacific, prime solid, diamond, star seeds. Just like the spacific sequence of the "five" spacific continuously balancing motions of the earth. Just like the transformed specific sequence of every cosmic whole componants "five" specific atmospheres, or "five" outermost epidermic spacific layers of "skin". Our universe continuously balances itself by continuously producing it's balancing components within the specific sequence of "five" specific gravity fields, or magnetic fields of continuous encouragement. This specific sequence of "five" specific gravity fields, or magnetic fields consists of: one spherical unified layer of our universe, the galcleous, the sun, the earth, and the moon. This continuous dynamic is how humans, and diamond is "constructed". There is only "**One**" specific cosmic whole component that is "The" constructor of matter. "The" regulator of matter. "The" balancer of matter. "The" continuous provider of the structural soundness of matter. This specific cosmic whole component has a specific sequence of "five" specific matter "constructors" protruding from it's core, or center. One head, two arms, and two legs. Each specific matter "constructor" has a specific sequence of "five" specific smaller matter "regulators". Each hand as "five" sequential fingers, and each foot has "five" sequential toes. Each head has a specific sequence of "five" specific transformed matter "balancers". Hearing, touching, smelling, seeing, and tasting. This specific cosmic whole component is also completely encapsulated, or "held together" or kept "structurally sound" by a spacific sequence of "five" outermost epidermic spacific layers of skin.

Matter produces one specific amount, and one specific sequence of transformations, or "patterns", or "graphs" of the amount of "five". "All" of them provide the exact same function for matter. This is the transformed pattern, or graph of "Every" constructor of matter. This is the transformed pattern, or graph of "Every" regulator of matter. This is the transformed pattern, or graph of "Every" balancer of matter. This is the transformed pattern, or graph of "Every" provider of structural soundness for matter.

This is the transformed pattern, or graph of "**YOU**".

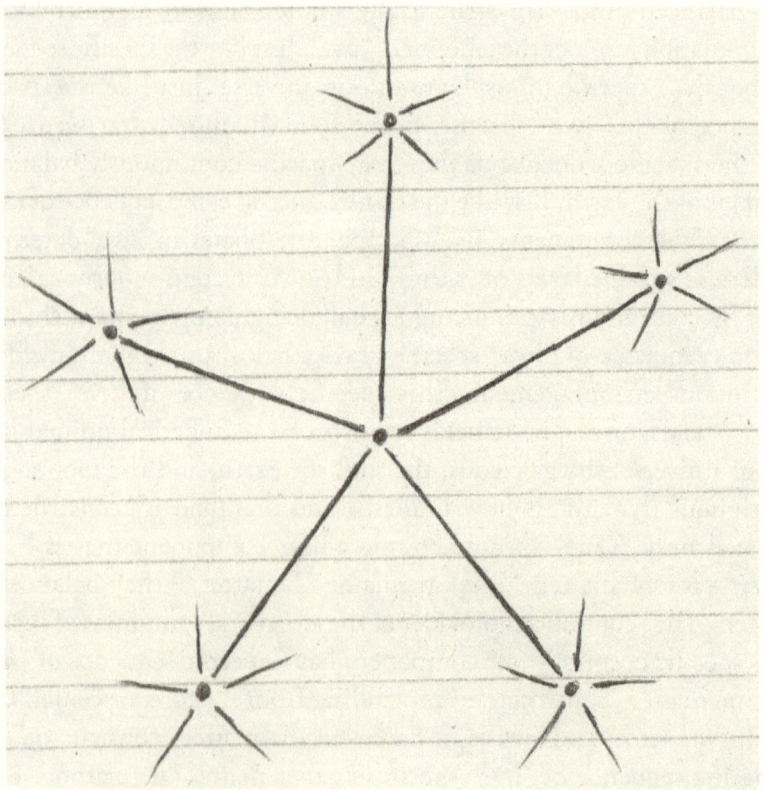

The continuous constructor, and regulator, and balancer, and provider of structural soundness for our solar system, and our galaxy, and our universe, and of matter itself is the advanced human like creature. The amount of "five" is the "middle", or "central" amount, and also the "external", or "final" amount, and also the "constructing" amount, and the "regulating" amount, and the "balancing" amount, and the amount that provides the "structural soundness" in the "key" continuously oscillating, continuous, repetitive, "stabilizing" specific amount, and specific sequence cycle of matter, centered around the "key" continuous, repetitive, "balancing" specific amount, and specific sequence cycle of matter. The amount of "five" is the continuous "constructor" of cosmic balance. The amount of "five" is the continuous "constructor" of matter itself.

Two

The amount of "two" is the "reproducing" amount, in the "key" continuously oscillating, continuous, repetitive, "stabilizing" specific amount, and specific sequence cycle of matter, centered around the "key" continuous, repetitive, "balancing" specific amount, and specific sequence cycle of matter.

Matter itself reveals that matter cannot produce "one" cosmic whole component, without there first being "two" specifically different cosmic whole components to produce it.

Matter itself has revealed which came first between the chicken, or the egg. The answer is neither one came first. They "Both" simply just keep coming. The amount of "two" is the reproducer of cosmic balance. The amount of "two" is the reproducer of matter itself.

This continuously balancing repetitive, specific amount sequence cycle, is the "key" to our universe's continuously balancing behavior.

365.25 and so on, and so on, and so on.....................

When you buy something that needs to be "assembled", they give you a set of "instructions". They give you a set of "directions". This set of directions is a specific amount, and a specific sequence of specific behaviors that are needed to produce one specific "whole", out of a specific amount, and a specific sequence of specific smaller "wholes".

Every cosmic whole component is produced from a specific, continuously balancing amount, and a specific, continuously balancing sequence of specific smaller cosmic whole components. Every specific cosmic whole component is a specific, continuously balancing amount, and a specific, continuously balancing sequence of specific, continuously balancing repetitive "sound cycles". Every specific sound, in every specific sound amount sequence cycle, is also a specific, continuously balancing amount, and a specific, continuously balancing sequence of specific, continuously balancing repetitive sound cycles, and so on, and so on, and so on.....................

In our universe, as well as in every other universe, there is "**one**" specific base amount, and "**one**" specific base sequence of specific, continuously balancing repetitive sound cycles. This is the "Blueprint" of matter's continuously balancing behaviors. This is the "Encyclopedia" of matter's continuously balancing behaviors. This is the "Dictionary" of matter's continuously balancing behaviors. This is the "Instructions" of matter's continuously balancing behaviors. This is the "Directions" of matter's continuously balancing behaviors. This is the Electro Magnetic Spectrum.

The Electromagnetic Spectrum

The electromagnetic spectrum is the directions of matter. It is the instructions of matter. Every specific cosmic whole accumulation of atomic matter exhibits many different patterns and behaviors. For these patterns and behaviors to be understood, there must be a template of constant patterns and behaviors to compare them to. This template of constant patterns and behaviers is the electromagnetic spectrum. The electromagnetic spectrum is a specific, continuously balancing amount, and a specific, continuously balancing sequence of specific, continuously balancing "frequencies". There are about 365.25 of these specific, continuously balancing "frequencies" that we can "see". These specific, continuously balancing frequencies, when uncompromised, never change there order, and never change their velocity. It's the same information over, and over, and over, and over, and over, and over, and over, and over, and over, and over, and over, and over, and over, and over, and over, and over, and over again.

"All" of these specific, continuously balancing frequencies are synchronized with the "key" frequency of our universe, and matter itself. This specific, continuously balancing "key" frequency of matter, is the peak velocity exhibited by a propagating photon, or the speed of light. Every specific, continuously balancing cosmic whole compilation of atomic matter is "synchronized" with the speed of light. Every specific, continuously balancing cosmic whole compilation of atomic matter exhibits a specific, continuously balancing synchronized frequency signature. "You" are a cosmic whole compilation of a specific, continuously balancing amount, and a specific, continuously balancing sequence of specific, continuously balancing atomic, and specific, continuously balancing organic molecular "frequencies". Your heart "beats" every time all of your specific, continuously balancing atomic and organic molecular frequencies are "synchronized" with the "key" specific, continuously balancing frequency of our universe. When you exert yourself for an extended period of time, your heart rate will increase, then very slowly decrease until it stabilizes. Even though your heart periodically speeds up and slows down a little, it still remains synchronized with the "key" specific, continuously balancing frequency of our universe, because it is very fast. As fast as the speed of light. Your heartbeat is your cosmic, continuously balancing synchronized frequency. Every spacific cosmic whole compilation exhibits one or more spacific, continuously balancing synchronized "frequencies".

Our sun's orbit around the center of our galaxy is a specific, continuously balancing synchronized frequency. Our sun's rotation is a specific, continuously balancing synchronized frequency. Our sun's flipping behavior is a specific, continuously balancing synchronized frequency. Our sun's wobbling behavior is a specific, continuously balancing synchronized frequency. Our sun's quaking behavior is a specific, continuously balancing synchronized frequency. Our earth's orbit around the center of the galaxy is a specific, continuously balancing synchronized frequency. Our earth's orbit around the sun is a specific, continuously balancing synchronized frequency. Our earth's rotation is a specific, continuously balancing synchronized frequency. Our earth's wobbling behavior is a specific, continuously balancing synchronized frequency. Our earth's flipping behavior is a specific, continuously balancing synchronized frequency. Our earth's quaking behavior is a specific, continuously balancing synchronized frequency. Every plant exhibits a specific, continuously balancing synchronized frequency. Every animal exhibits a specific, continuously balancing synchronized frequency. Every insect exhibits a specific, continuously balancing synchronized frequency. Every "Tone" is a specific, continuously balancing synchronized frequency. Every "Note" is a specific, continuously balancing synchronized frequency. Every "Pitch" is a specific, continuously balancing synchronized frequency. Every "Melody" is a specific, continuously balancing synchronized frequency. Every "Harmony" is a specific, continuously balancing synchronized frequency. Every "Rhythm" is a specific, continuously balancing synchronized frequency. Every "Tempo" is a specific, continuously balancing synchronized frequency. Every "Beat" is a specific, continuously balancing synchronized frequency. Every cellular compilation exhibits a specific, continuously balancing synchronized frequency. Every cell exhibits a specific, continuously balancing synchronized frequency. Every molecular compilation, and organic molecular compilation exhibits a specific, continuously balancing synchronized frequency. Every molecule, and organic molecule exhibits a specific, continuously balancing synchronized frequency. Every element exhibits a specific, continuously balancing synchronized frequency. Every atom exhibits a specific, continuously balancing synchronized frequency. Every proton, and every neutron, and every electron exhibits a specific, continuously balancing synchronized frequency. Every complete circle cycle of a propagating photon, is a specific, continuously balancing synchronized frequency. Every subatomic particle exhibits a specific, continuously balancing synchronized frequency. Every super subatomic particle exhibits a specific, continuously balancing synchronized frequency. Every super super subatomic particle exhibits a specific, continuously balancing synchronized frequency. Every super super super subatomic particle exhibits a specific, continuously balancing synchronized frequency, and so on, and so on, and so on.....................

Middle Matter

The cosmos is a continuously active continuum of endless larger encapsulating universes, and endless smaller encapsulated universes. There is no starting point of matter, and there is no end. Matter can only exist as a continuous encouragement, "**And**" a continuous response dynamic. It's the only way that matter can continuously exhibit a set formula of expectancy. It's the only way an electron, or a photon can continuously anticipate, and continuously respond to every variable they encounter in our universe. The electron, and the photon reveal that matter does not observe a scale of size, or velocity. There is no such thing as too slow, or too fast, or too large, or too small. The electron, and the photon exhibit perpetual momentum, and perpetual stability. The only way this is possible is with an endless availability of continuous internal information. The Deep Torsion Constant is the only way an electron, or a photon can exhibit a perpetually balancing behavior indefinitely. The photon, and the human reveal that matter cannot exist without the continuous encouragement from larger and slower accumulations of matter, "**And**" the continuous calculation, and the continuous response from smaller and faster accumulations of matter. There is no particle of matter that is a singularity. In the cosmos, everything is "middle" matter. "All" matter is in the "middle" of matter's behavior. Every cosmic whole component is in the "middle" of a larger and slower cosmic whole component. Every universe is in the "middle" of a larger encapsulating universe. "Exactly" in the "middle".

"**All**" matter is "**Middle**" matter.

The cosmic continuum continuously, rhythmically "pumps" larger and slower accumulations of matter that are "filled" with information outward, and continuously, rhythmically "sucks" smaller and faster accumulations of matter with less information inward, and our universe is right in the "**Middle**".

Now

The atomic matter that fills our universe is "fantastic". The subatomic matter that makes up our universe's atomic matter is even more "fantastic". Our universe's behavior is "constant". Every universe's behavior is "constant". In the cosmic continuum, everything is "continuously" happening. There is no such thing as traveling into your future. There is no such thing as traveling into your past. These things do not exist. Matter is "fantastic", but it cannot produce your future, or your past. Matter can only produce "now".

When we look at one of your atoms, we see that all of it's continuous behaviors are happening:

"Now".

All of an insect's continuous behaviors are happening "now". All of our moon's continuous behaviors are happening "now". All of our Earth's continuous behaviors are happening "now". All of our solar system's continuous behaviors are happening "now". All the continuous behaviors of our galaxy are happening "now". All of the continuous behaviors of our universe are happening "now". All of the continuous behaviors of the cosmic continuum are happening "now". The atom, and the galaxy reveal that everything is continuously happening, and "everything" is "continuously" happening:

"Now".

Time

We see biological organisms live and die every day. This is simply the activation, and the deactivation of a specific cosmic whole compilation of atomic matter. When a cosmic whole compilation of atomic matter deactivates, it disintegrates into smaller cosmic whole compilations of atomic matter. When an atom deactivates, it disintegrates into protons, electrons, and neutrons. When you look at your "hand", you are looking at many protons, and electrons, and neutrons. That would make the hand you are looking at, which is "your" hand, at least thirteen billion years old. It is possible that the protons, and the electrons, and the neutrons that are presently your hand, are even older than that. When you look at your hand, you're looking at subatomic particles that are much smaller, and much faster, and much "**older**" than your atomic particles. When you look at your hand, you're looking at super subatomic particles that are much smaller, and much faster, and much "**older**" than your subatomic particles. When you look at your hand, you're looking at super super subatomic particles that are much smaller, and much faster, and much "**older**" than your super subatomic particles and so on, and so on, and so on.....................

Our concept of time has no place when we contemplate the "age" of your super subatomic particles. Our concept of time has no place when we contemplate the "life span" of your super subatomic particles.

Matter itself reveals that it can "never" break down into unusable components. When atomic matter disintegrates into nothing more than subatomic neutrons, they simply return to the subatomic universe to be reassembled, and sent back up the ladder of matter.

The very small and very fast super subatomic particles that make up your hand, and every other part of you, and every other part of everything else in our universe reveal that the behavior of matter itself is "constant", and the age of matter itself is "constant". Matter does not observe a scale of "age". Matter does not observe a scale of "time". Matter itself is "Timeless".

On earth, we measure time by using the continuously balancing behaviors of our earth as a template. This is "not" a measurement of "time". This is a measurement of cosmic "timing". In the cosmic continuum of "endless" smaller encapsulated universes, and "endless" larger encapsulating universes,

there is no such thing as **"Time"**.

There is only continuously balancing synchronized cosmic **"Timing"**.

Our Universe

The center of our universe, is the nucleus of our universe. Every universe has a nucleus at it's center. Every universe is the nucleus of a larger encapsulating universe. Every universe encapsulates a smaller universe at it's center. Our universe encapsulates and encourages a never ending sequence of smaller spherical universes. Our universe is encapsulated and encouraged by a never ending sequence of larger spherical universes. The behavior of every universe is "constant". The age of every universe is "constant".

Every universe exhibits a continuously balancing "Big Bang" rate. Every universe's continuously balancing "Big Bang" rate is synchronized. The larger ones are slower by a specific multiple of the smaller ones, and the smaller ones are faster by a specific division of the larger ones. The continuously balancing "Big Bang" rate, or the univation rate of the subatomic universe, is more than every thirteen billion years. The continuously balancing Big Bang rate, or the univation rate of the super subatomic universe, is the same as your continuously balancing heart rate. "Bang", "Bang", "Bang", "Bang". These continuously balancing Big Bangs, or univations provide the protons, and the neutrons, and the electrons for every universe's continuous behavior.

The continuously balancing behavior of every universe, is the continuously balancing production of the neutron for the next advancing larger universe. This is continuously accomplished by advanced human like creatures.

The electromagnetic spectrum is the "Directions" of matter. It is the blueprint for everything there is in our universe, and how to assemble it.

Advanced human like creatures translate this information into "**sound**", then they transpose this information into our sun, and our solar system, and every other solar system, and every galaxy in our universe.

The advanced humans begin the production of the post atomic universe's nutrons by seeding the wave of Big Bang particles, or univation particles, with galaxy activating star "seeds". Every galaxy activating star seed is the same. It is a "spinning" tetrahedron diamond. A star like our sun is seeded with a "spinning" icosahedron diamond. A star that is one and a half times the mass of our sun, is seeded with a "spinning" dodecahedron diamond. A star that is twice the mass of our sun, is seeded with a "spinning" octahedron diamond, and a star that is three times the mass of our sun, is seeded with a "spinning" cube diamond.

Every planet, and every moon in our solar system, and in our galaxy, and in our universe, is "seeded" by advanced humans. They are "all" continuously cultured and guided, with comets and asteroids.

Every galaxy in our universe is being "continuously" guided to it's next destination. Our galaxy is "continuously" being guided toward the Andromeda galaxy by advanced humans. As our galaxy travels through space, it spins like a gyroscope. As our galaxy's nucleus continuously "eats" the stars and planets that are continuously "attracted" toward the nucleus, advanced humans continuously produce new torquing components. A continuously balancing introduction of new "torquing" stars and planets is how our galaxy, and every other galaxy is "continuously" "steered", or "guided" to it's next intended destination. "Every" star in our universe is "seeded" by advanced humans. "Every" planet, and moon is "seeded" by advanced humans. "Every" planet, and moon is "continuously" cultured and guided, with comets and asteroids by advanced humans.

To reach, and continuously maintain the absolute highest level of evolution that a human like creature can achieve, "All" advanced humans use the exact same formula.

The application of fact and reson "must" always temper emotion, but emotion "must never" be tempered by the application of universal fact and reson.

The result of this equation, or formula is:

"Continuously balancing" maximum Health.
"Continuously balancing" maximum Strength.
"Continuously balancing" maximum Wisdom.

Advanced humans maintain the continuously balancing maximum evolutionary level of "Health", "Strength", and "Wisdom". Advanced humans maintain the continuously balancing maximum evolutionary level of health, strength, and wisdom by continuously "caring".

The advanced humans in our universe, and every smaller and larger universe continuously "care" differently than today's Earth humans. Advanced humans culture a continuously balancing environment to maintain the maximum level of health, and strength. This is the maximum level of cosmic wisdom. Advanced humans continuously produce replicants with strong teeth. They do not continuously produce replicants to become dentists. Advanced humans continuously exhibit a behavior that continuously maintains the highest possible amount of health, and strength. Advanced humans do not continuously exhibit a behavior that continuously maintains the highest possible amount of hospital patients. The continuously balancing successful patterns and behaviors of "nature" is the "template" for "all" of the advanced humans in our galaxy, and every other galaxy in our universe, and every other universe in the cosmic continuum.

The continuously balancing behavior of the ucleous, or the nucleus of our universe, or the subatomic universe "incubates" the continuously balancing behavior of the nucleus of our galaxy. The continuously balancing behavior of the nucleus of our universe, and the nucleus of our galaxy incubates the continuously balancing behavior of the stars in our galaxy. The continuously balancing behavior of the nucleus of our universe, and the nucleus of our galaxy, and the star that is our sun incubates the continuously balancing behavior of our earth, and moon. The continuously balancing behavior of the nucleus of our universe, and the nucleus of our galaxy, and the nucleus of our solar system, and our earth, and our moon incubates the continuously balancing production of humans, and diamond.

The diamonds and the humans that every earth continuously produces are used to build the "machines" that "produce" every galaxy seed, and every star seed, and every planet seed, and every moon seed.

The diamonds and the humans that every earth continuously produces are used to build the "machines" that "apply" every galaxy seed, and every star seed, and every planet seed, and every moon seed.

The "seed" of a planet or moon does not occur naturally. Planet seeds, and moon seeds "must" be produced by advanced humans.

There is "**No**" random planetary behavior in the universe.

The "diamonds" that make up every star "seed", and every galaxy activating superstar "seed", are "not" produced by the Earth naturally. They are produced by advanced humans. Advanced humans produce the diamonds that make up every star seed, and every galaxy activating super star seed, by first culturing a continuously balancing environment of maximum health, and strength. In this continuously balancing environment, advanced humans "eat" an uncompromised specific, continuously balancing amount, and specific, continuously balancing sequence of specific, diamond, star seed base components. When the advanced human has accumulated, and fully processed a specific, continuously balancing amount, and a specific, continuously balancing sequence of this specific, continuously balancing diet, the advanced human, along with a specific, continuously balancing amount, and a specific, continuously balancing sequence of other specific advanced humans are "**themselves**" processed into the final specific, diamond, star seed composite.

Advanced humans are themselves the formative composite for every specific diamond, star seed, and galaxy activating super star seed.

Advanced humans "**Are**" the star seeds.

Advanced humans regulate "every" formative phase of a diamond, star seed's production, and "every" phase of it's transportation, and "every" phase of it's application. In our universe, as in every smaller universe, and in every larger universe, the "**random**" star producing behavior is 0.0%.

Every atomic particle that "**you**" are made up of, is itself made up of "many" fully processed, subatomic, super diamond, star, and super star seeds. All of these subatomic, fully processed, super diamond, star seeds are made up of countless advanced subatomic human like creatures, from the subatomic universe

"**You**", and "every" other atomic cosmic whole component, are made up of "countless" subatomic advanced humans, and "countless" super subatomic advanced humans, and "countless" super super subatomic advanced humans, and "countless" super super super subatomic advanced humans, and so on, and so on, and so on.....................

Every universe is one complete, continuously balancing accumulation transformation sequence cycle of matter. Every universe is one complete, continuously balancing, repetitive sequence cycle of cosmic information. Every universe is made up of a specific, continuously balancing amount, and a specific, continuously balancing sequence of specific unified layers.

Every unified layer of our universe, and every other universe, is a spherical wave that encapsulates a smaller spherical wave, and is also encapsulated by an even larger spherical wave.

Every specific unified layer of every universe in the cosmic continuum, is made up of a specific, continuously balancing amount, and a specific, continuously balancing sequence of specific unified particle "clusters". Each specific unified particle "cluster" orbits a specific larger, central, spherical unified particle "cluster" layer, made up of a specific, continuously balancing amount, and a specific, continuously balancing sequence of even "smaller" specific unified particle "clusters", and so on, and son, and so on.....................

Every unified layer in our universe, and every unified layer in every other universe, is the exact same information as the smaller encapsulated one before it, and the larger encapsulating one after it. Every unified layer of the cosmic continuum is the exact same information as each, and every other layer. Each specific unified layer of the cosmic continuum is simply a specific accumulation transformation of each, and every other layer. Every unified layer of the cosmic continuum is continuously interconnected, and continuously interactive.

The "Continuously balancing" behaviors of our universe are revealed as we travel to the center of our universe, starting from our solar system.

Our "solar system" is in the "unified active galaxy cluster", unified layer of our universe.

Our "solar system" is made up of a unified "cluster" of rotating, wobbling, flipping, and quaking unified "clusters", which are planets and moons, orbiting a larger, central, unified "cluster" of even smaller unified "clusters". Which is the sun.

Our "sun" is a rotating, wobbling, flipping, and quaking star that is part of a rotating, wobbling, flipping, and quaking unified "cluster" of stars that orbit a larger, central, unified "cluster" of even smaller unified "clusters", which is the "nucleus" of our galaxy.

Our "galaxy" is a rotating, wobbling, flipping, and quaking unified "cluster" of stars, and planets, and moons, that is part of a rotating, wobbling, flipping, and quaking unified "cluster" of galaxies that orbit a larger, central, unified "cluster" of even smaller unified "clusters", which is the unified spherical layer of the universe that precedes the "unified active galaxy cluster", unified spherical layer of the universe.

This specific dynamic now leads back to the subatomic universe.

The central unified "cluster" that every "unified active galaxy cluster" orbits, is a unified spherical layer of the universe. This unified layer is much larger than each of the orbiting "unified active galaxy clusters", and each of it's unified "clusters" are much smaller than each of the "unified active galaxy clusters".

This is our Universe. Our unified active galaxy cluster is here ⟶ ⊙

Each rotating, wobbling, flipping, and quaking "unified active galaxy cluster" orbits a larger, central, spherical layer of "smaller" rotating, wobbling, flipping, and quaking unified "clusters" of radiant galaxy centers, or "super stars".

Each rotating, wobbling, flipping, and quaking unified radiant super star "cluster", orbits a larger, central, spherical unified layer of "smaller" rotating, wobbling, flipping, and quaking unified "clusters".

These are "neutrons". This unified spherical layer of neutrons is the "Univator".

Each rotating, wobbling, flipping, and quaking unified neutron orbits a larger, central, spherical layer of "smaller" rotating, wobbling, flipping, and quaking unified "clusters". These smaller unified "clusters" are incomplete neutrons. This unified spherical layer of rotating, wobbling, flipping, and quaking unified incomplete neutrons, is the "outside" of the subatomic universe. It is the "Fifth" exterior layer of the subatomic universe. It is the "Fifth", and outermost uniousphere, of the subatomic universe's "five" sequential uniouspheres. The "five" sequential, external, spherical, unified "layers", or "waves", or uniouspheres of the subatomic universe "continuously" construct, and "continuously" regulate, and "continuously" balance, and "continuously" maintain the structural soundness of the subatomic universe, and "continuously" construct, and "continuously" regulate, and "continuously" balance, and "continuously" maintain the structural soundness of every atomic neutron. The continuous behavior of our universe, is the identical behavior of every smaller encapsulated universe, and every larger encapsulating universe in the cosmic continuum.

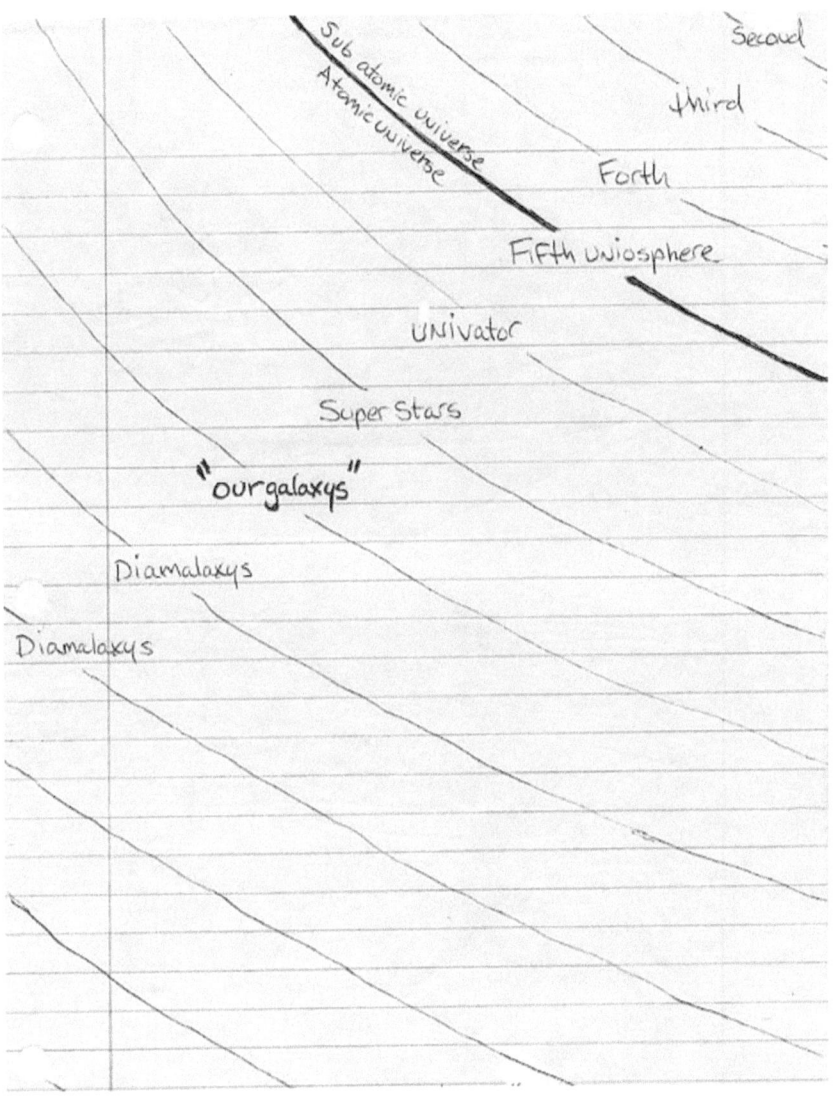

Sub atomic universe
Atomic universe
Second
third
Forth
Fifth uniosphere
univator
Super Stars
"ourgalaxys"
Diamalaxys
Diamalaxys

Every unified layer of every universe is the "Exact" same information. It is simply transformed.

Every universe is "one" specific, continuously balancing amount, and "one" specific, continuously balancing sequence of continuously interconnected, and continuously interactive, spherical unified layers. Every complete, spherical unified layer, of every universe's complete, spherical unified layer amount and sequence, is the "exact" same information. Each one is simply a transformed accumulation motion result.

The Deep Torsion Constant

Everything that the subatomic universe produces, is inside of every atomic neutron. Every atomic neutron is one specific, continuously balancing complete amount, and one specific, continuously balancing complete sequence of subatomic information, encapsulating one specific, continuously balancing complete amount, and one specific, continuously balancing complete sequence of the "Deep Torsion Constant". When the atomic neutron is separated, all of these specific separations behave independently. Each specific separation of the atomic neutron is represented by every photon, and every electron, and every proton in our universe.

"Every" proton, and "Every" electron, and "Every" photon is simply one specific, continuously balancing "**incomplete**" amount, and one specific, continuously balancing "**incomplete**" sequence of subatomic information, and one specific, continuously balancing "**incomplete**" amount, and one specific, continuously balancing "**incomplete**" sequence of the Deep Torsion Constant.

"Every" neutron, and "Every" proton, and "Every" electron, and "Every" photon is the "exact" same information. Each one is one specific, continuously balancing amount "**more**", or one specific, continuously balancing amount "**less**" of the "exact" same information. Each specific, continuously balancing amount of this information produces a different, and specific sequence of information, producing each individual atomic particle's specific, and different, and transformed overall "continuously balancing" motion result.

"Every" neutron, and "Every" proton, and "Every" electron, and

"Every" photon, is a specific, continuously balancing amount, and a specific, continuously balancing sequence of specific, continuously balancing unified subatomic layers, of specific, continuously balancing unified subatomic particle clusters, of specific, continuously balancing unified subatomic particles.

"Every" specific, continuously balancing complete or incomplete unified subatomic particle cluster layer amount, and sequence, is the "exact" same information.

That's Because;

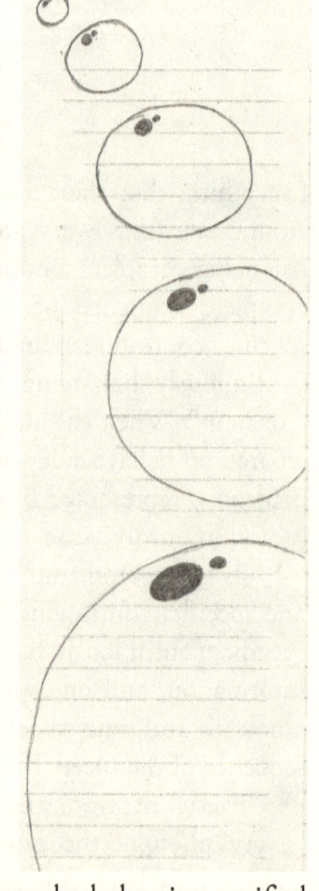

"Every" specific, continuously balancing complete, or incomplete, rotating, wobbling, flipping, quaking, and orbiting unified subatomic particle cluster, in every specific, continuously balancing complete or incomplete amount and sequence of unified subatomic particle cluster layers, is the "exact" same information.

That's Because;

"Every" specific, continuously balancing unified subatomic particle, in every specific, continuously balancing complete, or incomplete unified subatomic particle cluster, in every specific, continuously balancing complete, or incomplete amount and sequence of unified subatomic particle cluster layers, is the "**exact**" same specific, continuously balancing rotating, wobbling, flipping, quaking, and orbiting unified subatomic particle. Of course, this identical specific, continuously balancing unified subatomic particle is itself a unified particle cluster. "Every" identical specific, continuously balancing unified subatomic particle, is one "complete" super diamalaxy.

"Every" identical unified subatomic particle is comprised of one specific, continuously balancing amount, and one specific, continuously balancing sequence, of "five" specifically different, rotating, wobbling, flipping, quaking, and orbiting, fully processed, super compressed diamond, subatomic star seeds, and super star seeds. This unified particle of fully processed, super compressed diamond, subatomic star, and super star seeds "**Can**" disintegrate, but it does "**Not**" separate when the atomic neutron separates into photons, and electrons, and protons. This unified particle of a specific, continuously balancing amount, and a specific, continuously balancing sequence, of "Five" specifically different, fully processed, super compressed diamond, subatomic star, and super star seeds remains "complete" inside every complete, or incomplete, specific, continuously balancing amount and sequence of subatomic information.

Only the gyroscopic torsion of the continuously interactive, super diamond, subatomic star seeds changes. Each one of the "Five" fully processed, super diamond, subatomic star seeds is a different size. Each one is a specific amount larger, and, or a specific amount smaller than each of the other ones.

Each unified particle, of super diamond, subatomic star seeds, is a specific sequence, of "Five" specific unified layers.

The central unified first layer, is a specific, continuously balancing amount, and a specific, continuously balancing sequence of the smallest super diamond, subatomic star seeds, which are the fully processed seeds of the subatomic super stars, or galaxy activating stars.

The second, and next larger unified layer is a specific, continuously balancing amount, and a specific, continuously balancing sequence of the second largest super diamond, subatomic star seeds, which are the fully processed seeds of the subatomic stars that are three times the mass of a subatomic "sun".

The third, and next larger, and "middle" unified layer is a specific, continuously balancing amount, and a specific, continuously balancing sequence of the third largest, or middle sized super diamond, subatomic star seeds, which are the fully processed seeds of the subatomic stars that are twice the mass of a subatomic "sun".

The fourth, and next larger unified layer is a specific, continuously balancing amount, and a specific, continuously balancing sequence of the fourth largest super diamond, subatomic star seeds, which are the fully processed seeds of the subatomic stars that are one and a half times the mass of a subatomic "sun".

The fifth, and largest, and exterior, and last unified layer is a specific, continuously balancing amount, and a specific, continuously balancing sequence of the fifth largest super diamond, subatomic star seeds, which are the fully processed seeds of the subatomic "suns", which are identical to our "sun", only a specific amount smaller.

Each one of the "**five**" specifically different sized, fully processed, super diamond, subatomic star seeds is: five exactly the same continuously balancing amounts, and five exactly the same, but specifically different sized continuously balancing sequences of, or five specifically different continuously balancing amounts, and five specifically different continuously balancing sequences of the "exact same sized" continuously balancing rotating, wobbling, flipping, quaking, and orbiting subatomic neutrons, traveling at the exact same continuously balancing velocity, or five specifically different continuously balancing velocities.

"Only" smaller neutrons can travel through a universe that is not their original universe.

"Only" atomic neutrons, and neutrons that are smaller than atomic neutrons, can penetrate our universe's "Five" sequential encapsulating uniouspheres.

The "**SPACE**" of our universe is made up of subatomic neutrons, and super subatomic neutrons, and super super subatomic neutrons, and super super super subatomic neutrons, and super super super super subatomic neutrons, and super super super super super subatomic neutrons, and super super super super super super subatomic neutrons, and so on, and so on, and so on and so on, and so on, and so on, and so on, and so on, and so on, and so on, and so on....................

Every atomic neutron, and proton, and electron, and photon does "**not**" continuously "move" through "**Space**". They continuously "appear" and continuously "disappear", by continuously "accumulating", and continuously "disintegrating". Every atomic neutron, and proton, and electron, and photon continuously "accumulates" and continuously "disintegrates" in any direction, and in every direction, by continuously, in a transformed way, "incorporating", and "absorbing", and "inhaling", and "eating" a specific, continuously balancing amount, and a specific, continuously balancing sequence of subatomic neutrons, and continuously, in a transformed way, "exfoliating", and "transpiring", and "exhaling", and "excreting" a specific, continuously balancing amount, and a specific, continuously balancing sequence of subatomic neutrons, at a specific, continuously balancing amount of times "faster" than the speed of light.

"Every" subatomic neutron is a specific, continuously balancing amount, and a specific, continuously balancing sequence of super subatomic neutrons.

"Every" super subatomic neutron is a specific, continuously balancing amount, and a specific, continuously balancing sequence of super super subatomic neutrons.

"Every" super super subatomic neutron is a specific, continuously balancing amount, and a specific, continuously balancing sequence of super super super subatomic neutrons.

"Every" super super super subatomic neutron is a specific, continuously balancing amount, and a specific, continuously balancing sequence of super super super super subatomic neutrons.

"Every" super super super super subatomic neutron is a specific, continuously balancing amount, and a specific, continuously balancing sequence of super super super super super subatomic neutrons, and so on, and so on, and so on...................

This Is the "Deep Torsion Constant".

Matter itself reveals that matter is "**One**" specific, continuously balancing amount, and "**One**" specific, continuously balancing sequence cycle continuum, of accumulation transformations of "**One**" specific, continuously balancing particle, "**One**" specific, continuously balancing sound, "**One**" specific, continuously balancing system. "**Continuously Balancing Information Circulation**". Every specific transformation of continuously balancing information circulation, is produced by "**One**" specific, continuously balancing amount, and "**One**" specific, continuously balancing sequence of accumulation transformations, of "**Five**" specific, continuously balancing motions, **Rotating**, **Wobbling**, **Flipping**, **Quaking**, and **Orbiting**. Each one of these specific, continuously balancing motions are produced by "**One**" specific, continuously balancing amount, and "**One**" specific, continuously balancing sequence, of specific, continuously balancing accumulation transformations, of "**One**" specific, continuously balancing motion. This "**One**" specific, continuously balancing motion is "**Out Half Around Rest. Half Around Back Rest**".

Matter itself reveals that matter is **"One"** specific, continuously balancing amount, and **"One"** specific, continuously balancing sequence cycle continuum, of specific, continuously balancing accumulation transformations, of **"One"** specific, continuously balancing **"Journey"**.

The Journey of Matter.

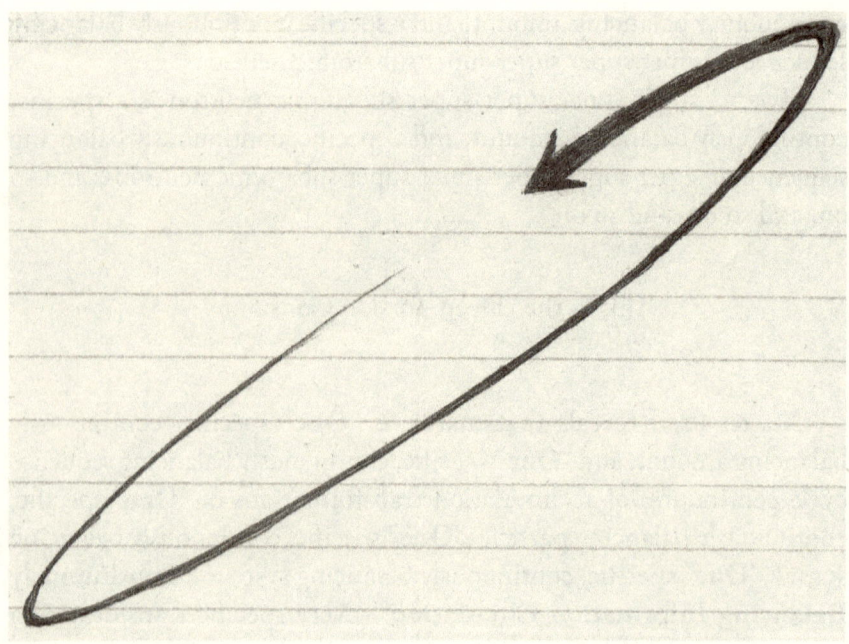

Glossary

*Deep Torsion Constant—Endless layers of smaller and smaller "Identical" information, produced by torquing components, producing matter's memory.

*Galactivation—The collapse, and explosion of a super star. An ultra nova, producing a galaxy center.

*Galactivator—It is a galaxy activator. It is an atomic particle accumulation regulating tetrahedron diamond. It is a galaxy seed.

*Galcleous—The center of every galaxy. It is a unified cluster of fully processed, super compressed diamond, star, and super star seeds.

*Super Diamalaxsphere—One specific, continuously balancing complete, or one specific, continuously balancing incomplete amount, and one specific, continuously balancing complete, or one specific, continuously balancing incomplete sequence, of unified layers, of unified particle clusters, of unified particles, of fully processed, super compressed diamond, subatomic star, and super star seeds. (Neutrons, Protons, Electrons and Photons.)

*Super Diamalaxy—One specific, continuously balancing complete amount, and one specific, continuously balancing complete sequence of fully processed, super compressed diamond, subatomic star, and super star seeds.

*Ucleous—(you'-klee-iss) The Center of the universe.

*Uniousphere—(you-nee'-o-sphere) One of the Five unified, sequential, external layers of the universe.

*<u>Univation</u>—The resulting explosion after the collapse of the univator. The "Big Bang".

*<u>Univator</u>—The unified layer of neutrons that encapsulates the subatomic universe.

Special thanks to:

Roy Sklarski

www.ingramcontent.com/pod-product-compliance
Lightning Source LLC
Chambersburg PA
CBHW021941170526
45157CB00003B/881